SpringerBriefs in Finance

For further volumes:
http://www.springer.com/series/10282

Manuela Spangler · Ralf Werner

German Covered Bonds

Overview and Risk Analysis of Pfandbriefe

Springer

Manuela Spangler
Allianz Deutschland AG
Munich
Germany

Ralf Werner
Institut für Mathematik
Universität Augsburg
Augsburg
Germany

The views and opinions expressed in this book are those of the authors only, and not those of their employers, Allianz Deutschland AG or DEVnet AG, their subsidiaries or affiliate companies. Nothing in this monograph is or should be taken as information regarding, or a description of, any operations, risk appraisal or consulting business of Allianz Deutschland AG or DEVnet AG.

ISSN 2193-1720 ISSN 2193-1739 (electronic)
ISBN 978-3-319-02552-0 ISBN 978-3-319-02553-7 (eBook)
DOI 10.1007/978-3-319-02553-7
Springer Cham Heidelberg New York Dordrecht London

Library of Congress Control Number: 2013953233

Printed on acid-free paper

Springer is part of Springer Science+Business Media (www.springer.com)

Preface

The long-lasting financial crisis at the end of the past decade—a period of severe market distress—has revealed that events that had been classified as rather improbable some years earlier can actually occur. Even German covered bonds (Pfandbriefe), which used to be considered as nearly risk-free and which had not incurred a single default in more than 200 years, were in some cases close to cover pool separation from the issuing bank, sometimes only avoided through bailouts of the corresponding issuers. Starting from there, we got interested in building a mathematical valuation model to assess the actual likelihood and severity of such events.

When starting our research we realized that there were almost no (academic) publications on this asset class: we could find very few detailed material on the Pfandbrief's cash flow mechanics, its risk profile or potential valuation methodologies. Furthermore, the majority of available information either came from lobbying and interest groups like vdp or ECBC, from rating agencies or from research departments of large investment banks, and was (assumably, at least to a certain degree) biased. This is when we decided that, before coming up with a proper mathematical valuation model, we have to lay the grounds for an in-depth understanding of the Pfandbrief mechanics.

Shortly after we had started our investigations into the Pfandbrief we were, by sheer coincidence, approached by a large insurance company, which was also looking for a detailed risk analysis of covered bonds. The risk managers there seemed to observe the same difficulties as we did in finding appropriate independent sources of information. Therefore, we suggested to provide a detailed introduction to Pfandbriefe, with a special emphasis on the legal framework together with a neutral risk analysis from the perspective of an investor.

While keeping our work up-to-date over the last three years, we have additionally collected a large amount of material covering different aspects of the Pfandbrief. This monograph now provides an extensive list of more than 200 references dealing with Pfandbriefe or covered bonds in general, which is why it is, to our opinion, an ideal starting point for further studies. Although focused on German legislation it can also provide valuable insights when dealing with other legal or contractual frameworks. We believe that—especially in the light of the growing number of countries with covered bond legislations and discussions about SME-backed covered bond issues in Europe—it might provide additional benefits to some readers.

When writing this monograph, we had in mind the following target audience: risk managers at large financial institutions with significant investments in German Pfandbriefe (like insurance companies, pension funds and asset management companies), rating agencies and regulators. Although not in the main focus, it might be equally valuable for employees in the banking industry who are concerned with Pfandbrief-related duties such as issuance, pricing, reporting and risk management. We hope that the monograph will also be welcomed by students in banking and finance looking for concise and complete information on the German Pfandbrief, and that it might act as a starting point for fellow research colleagues in new pricing models for Pfandbriefe.

As we plan to continue our research in Pfandbriefe, we welcome any constructive feedback on this monograph, e.g. additional content to be included in a potential further edition, different opinions on some subjects or errors which need to be corrected.

Munich, August 2013

Manuela Spangler
Ralf Werner

Acknowledgments

This monograph is based on the experience gained from several years of working in an international real estate financing bank, and on insights obtained during projects with a large insurance company. As such, it has been influenced by many people to whom we would like to express our gratitude.

First of all, we would like to thank our former colleagues for the good cooperation over the years. Some of them have contributed a lot to our understanding of the covered bond business. Special thanks in this context go to Andreas Schenk, for his helpful support, and to Werner Neumeier, for many fruitful discussions. Sincere thanks go to Larisa Angstenberger, Robert Wania, Martin Fischer, Bernhard Held and Pierre Joos for initiating the risk analysis of German covered bonds and their friendly cooperation during the projects. We would also like to express our gratitude to the colleagues at DEVnet, especially Nikolaj Leonhardt, for his support during the publishing phase of this book, and Andi Kokott, for the general support of the project. We are also thankful to Selma Uhlig (Deutsche Pfandbriefbank), Yan Yang and Max Hughes (both from Hochschule München) for their numerous helpful comments on a previous version of our manuscript. Last but not least, we would like to thank the European Covered Bond Council (ECBC) and Bernd Volk (Deutsche Bank) for their consent to reference some of the statistics and figures from their publications.

Contents

German Covered Bonds

Abstract With a history that goes back to the eighteenth century and a high market share in today's covered bond markets, the German Pfandbrief is the most established covered bond. Until today, no single Pfandbrief has ever defaulted. Even though Pfandbriefe have survived the financial crisis comparably unharmed, investors have become more sensitive regarding the creditworthiness of the corresponding issuer and sovereign, the strength of the legal (or contractual) framework and the quality of the cover pool serving as collateral. This monograph provides a structured in-depth analysis of the legal framework and the risks inherent in a Pfandbrief, taking into consideration recent market developments. Starting from the legal framework, the German Pfandbrief is introduced without requiring prior knowledge. Covered bond related risks are explained in detail and their relevance to the Pfandbrief is thoroughly discussed with focus on the two most common Pfandbrief types, Mortgage and Public Pfandbriefe. In addition to that, the monograph comes with an extensive collection of Pfandbrief-related literature and a glossary explaining the main technical terms.

Keywords Pfandbrief · Covered bond · Structured covered bond · Cover pool · Mortgage covered bond · Public sector covered bond · Financial risk analysis · Legal framework · Cover requirements

1 Introduction

With a history that goes back to the eighteenth century and a high market share in today's covered bond markets, the German Pfandbrief is the most established *covered bond*. Until today, no single Pfandbrief has ever defaulted, and the Pfandbrief is one of few financial instruments that have survived the financial crisis comparably unharmed. Due to its long track record and its high quality standards, the Pfandbrief has always been very popular with investors. Like other covered bonds, it also benefits from privileged treatment in ECB liquidity operations and in various areas of

M. Spangler and R. Werner, *German Covered Bonds*,
SpringerBriefs in Finance, DOI: 10.1007/978-3-319-02553-7_1,

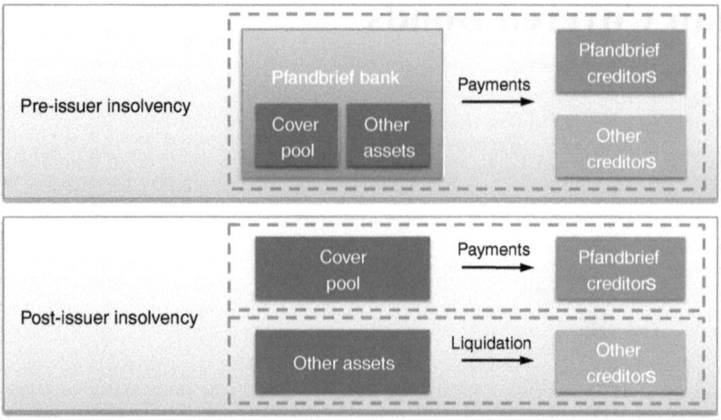

Fig. 1 The Pfandbrief's dual protection mechanism

EU financial market regulation, such as Basel III/CRD IV (e.g. liquidity coverage requirements) and Solvency II (e.g. capital requirements).

Before the financial crisis, Pfandbriefe were considered to be nearly risk-free, and their spreads to government bonds were often interpreted as a mere liquidity premium. Investors used to rely on the Pfandbrief's AAA rating without conducting their own analysis, but this has now changed: Pfandbrief investors and, more generally, covered bond investors have become more sensitive regarding the creditworthiness of the corresponding issuer, the strength of the legal (or contractual) framework, and the quality of the cover pool serving as collateral. While in 2008 issuer risk was the major driver for spread differentiation, sovereign risk has been driving spread differences between covered bonds from different jurisdictions since 2010. The time of AAA ratings is over: it has become more and more difficult for issuers to maintain top Pfandbrief ratings, mainly caused by a decrease in issuer ratings and increased rating agency requirements.

A *Pfandbrief* is a covered bond issued under German Pfandbrief legislation. Its main feature is its dual nature of protection (see Fig. 1). The Pfandbrief holder has full recourse to the issuer, and, in case of issuer insolvency, he also has a preferential claim on a dedicated pool of assets, the cover pool. Cover pool assets (which can be either claims against public sector debtors or mortgages on real estate properties, ships or aircraft) remain on the issuer's balance sheet but are registered in the cover register. As long as the issuer is solvent, he is expected (and to a certain extent required) to maintain the quality and the size of the cover pool by replacing maturing, non-performing and non-eligible assets, or by posting additional collateral. Therefore, the cover pool is usually dynamic. Upon issuer insolvency, it is segregated from the issuer's general insolvency estate and administered until all Pfandbriefe are repaid according to their contractual terms. In case the proceeds from the cover pool are not sufficient to repay the Pfandbrief holders, the cover pool is liquidated, and the Pfandbrief holders' remaining claims rank *pari passu* with the unsecured creditors of the issuer. In contrast to senior unsecured debt, Pfandbriefe do not automatically become due upon issuer default, i.e. issuer default does not trigger Pfandbrief *acceleration*.

The purpose of this work is a structured in-depth analysis of the legal framework and the risks inherent in a German Pfandbrief from an investor's point of view, taking into consideration recent market developments. Starting from the legal framework, we introduce the German Pfandbrief, explain covered bond related risks discussed in the literature and investigate their relevance to the German Pfandbrief. We do so by focusing on the two most common Pfandbrief types in Germany: Mortgage and Public Pfandbriefe. In addition to that, we provide an extensive collection of Pfandbrief related literature in the References section. To our knowledge there is, so far, no such analysis particularly devoted to the German Pfandbrief. Although the Association of German Pfandbrief Banks (*vdp*) has published a huge amount of informative literature on the German Pfandbrief (most of which can be downloaded from the vdp's website vdp 2013k), none of these publications cover such an analysis. Rating agencies, on the contrary, have written a vast amount of covered bond related publications (see References section), but most of them deal with covered bond risks in general, and are not particularly devoted to the German Pfandbrief. One notable exception is Moody's (2010d), which summarizes the legal framework for Pfandbrief issuance and outlines the rating agency's main findings with respect to the strengths and weaknesses of the German Pfandbrief. This analysis was our first starting point.

The remainder of this work is structured as follows. In Sect. 2, we describe the legal framework for German Pfandbriefe, focusing on aspects that are relevant to the subsequent risk analysis. In Sect. 3, a brief overview of the covered bond market and the role the German Pfandbrief plays in this market can be found. Section 4 gives an extensive description of covered bond related risks and discusses their relevance to the German Pfandbrief. It is also explained to what extent these risks are mitigated by legal provisions, the German Pfandbrief system in general and by program-specific features. Finally, Sect. 5 concludes the risk analysis.

2 Legal Framework

This section gives an introduction to German Pfandbrief legislation with a focus on the features that are relevant to the subsequent risk analysis. It is mainly based on the vdp's collection of legislative materials vdp (2011d), and an article by Tolckmitt and Stöcker (2012b). For more details on legal aspects, Koppmann (2009) and vdp (2005c) are a good source of information.

2.1 Legislative Materials

As in most European jurisdictions, Pfandbrief issuance is based on a legislative framework (i.e. on a special law),

- the German Pfandbrief Act ("Pfandbriefgesetz", *PfandBG*), which was introduced in 2005 and amended in 2009 and 2010.

Furthermore, there are four additional regulations (statutory orders) published by the German Financial Supervisory Authority (*BaFin*) that are relevant to Mortgage and Public Pfandbriefe:

- the Net Present Value Regulation ("Pfandbrief-Barwertverordnung", *PfandBarwertV*), which specifies the calculation of the net present value to be used for the determination of cover requirements,
- the Regulation on the Determination of the Mortgage Lending Value ("Beleihungswertermittlungsverordnung", *BelWertV*),
- the Cover Register Statutory Order ("Deckungsregisterverordnung", *DeckRegV*), which regulates the content and the handling of the cover register, and
- the Funding Register Statutory Order ("Refinanzierungsregisterverordnung", *RefiRegV*), which regulates the content and the handling of the refinancing register according to § 22a to § 22o of the German Banking Act ("Kreditwesengesetz", *KWG*).

There are two more regulations concerning Ship and Aircraft Pfandbriefe which are, however, not relevant to the subsequent risk analysis:

- the Regulation on the Determination of the Mortgage Lending Value of Ships and Ships under Construction ("Schiffsbeleihungswertermittlungsverordnung", *SchiffsBelWertV*), and
- the Regulation on the Determination of the Mortgage Lending Value of Aircraft ("Flugzeugbeleihungswertermittlungsverordnung", *FlugBelWertV*).

When we refer to *Pfandbrief legislation* in the following, we mean all these legislative materials, which are illustrated by Fig. 2. Apart from the PfandBG, Pfandbrief banks are also subject to the German Banking Act, as they are credit institutions according to § 1 par. 1 KWG.

At the time of writing, another amendment of the PfandBG had already been passed: The legislative package that will transpose CRD IV into German law (see Bundesrat 2013) contains changes to the PfandBG which include, among others, increased § 28 transparency requirements, a revision of the rules governing the position of the cover pool administrator, and a more precise definition of related procedures. The legislative package is scheduled to take effect on January 1st, 2014.

2.2 *Pfandbrief Issuance*

The prerequisites for Pfandbrief issuance are specified by § 2 PfandBG. To engage in Pfandbrief business, a banking license from BaFin according to § 32 KWG is needed. Furthermore, the following requirements need to be fulfilled:

- A core capital of at least EUR 25 mn,

Fig. 2 Pfandbrief legislation

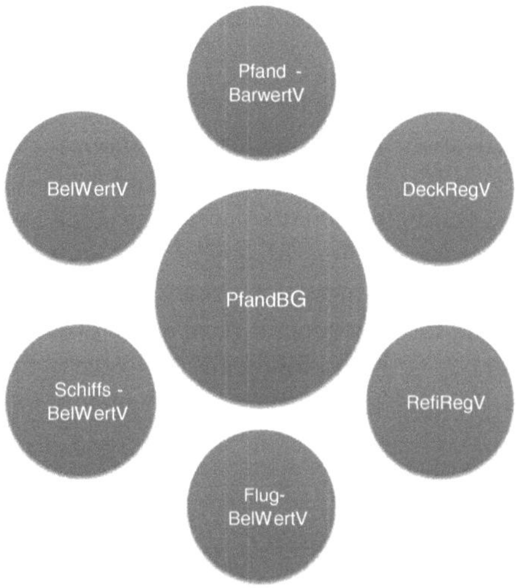

- a license to engage in lending business according to § 1 par. 1 sent. 2 no. 2 KWG,
- suitable procedures and instruments to manage, monitor and control the risks arising from the cover pools and the Pfandbrief issuance activities,
- a regular and sustainable Pfandbrief business plan, and
- a sufficient organizational structure to support mortgage and public sector lending including its refinancing via Pfandbriefe.

A bank holding a license for Pfandbrief issuance is called *Pfandbrief bank*. In the following, we also refer to the Pfandbrief bank as *Pfandbrief issuer*. Pfandbriefe can be used to fund loans secured by claims against public sector debtors (*Public Pfandbrief*) or mortgages on real estate properties (*Mortgage Pfandbrief*), ships (*Ship Pfandbrief*) or aircraft (*Aircraft Pfandbrief*). For each Pfandbrief type, a separate license is necessary. BaFin has the right to withdraw the Pfandbrief license when the above requirements are not fulfilled any more, or when the Pfandbrief bank has not issued new Pfandbriefe for more than two years and it is not expected that Pfandbrief business is going to be resumed on a regular and sustainable basis during the next six months. In these cases, the cover pool would be settled.

Before the PfandBG was introduced in 2005, a *specialist bank principle* was in place. German mortgage banks were restricted to specific business activities, which were considered to be "fundamentally safe", and non-eligible business was limited to 20 % of their mortgage lending activities (Volk 2011a, p. 20). Under today's *universal bank principle*, there are—apart from the (general) regulations of the German Banking Act—no restrictions regarding the business activities a Pfandbrief issuer is allowed to engage in.

2.3 Preferential Claim

Upon issuer insolvency, Pfandbrief holders and counterparties from cover pool derivatives have a preferential claim on the Pfandbrief's cover pool (§ 29 et seqq PfandBG). In case the Pfandbrief bank has issued more than one Pfandbrief type, the preferential claim of a specific Pfandbrief type refers to the particular cover pool only; there is no *cross-collateralization*. To ensure the preferential claim, the cover pool is ring-fenced, i.e. in case of issuer default it is isolated and does not participate in the issuer's general insolvency proceedings. This is specified in more detail in Sect. 2.12. In the following, we refer to Pfandbrief holders and counterparties from cover pool derivatives as *privileged creditors*, or *preferred creditors*.

2.4 Cover Register

The main purpose of the *cover register*, which is addressed by § 5 PfandBG, is to safeguard the preferential claim of the privileged creditors in case of issuer insolvency. To make cover assets and cover pool derivatives identifiable, they have to be registered in the respective cover register, and a separate cover register is maintained for each Pfandbrief type. For assets that only partially belong to the cover pool,[1] the extent to which they belong to the cover pool and the ranking of the cover pool part (in relation to the part not belonging to the cover pool) need to be specified, too. In case of doubt, the cover pool part has priority. Derivatives can only be registered with the corresponding counterparty's consent. Details regarding the maintenance of the cover register are specified in the DeckRegV. In the following, we refer to registered cover assets and derivatives as *cover pool*.

2.5 Eligibility Criteria

Ordinary Cover Assets

To be eligible for inclusion in cover ("deckungsstockfähig"), *ordinary cover assets* must fulfil certain *eligibility criteria* which mainly depend on the Pfandbrief type. In the following, these eligibility criteria are described in more detail.

[1] Residential and commercial mortgages, for example, can only be included in the cover pool up to 60 % of the mortgage lending value, see Sect. 2.5.

Mortgage Pfandbriefe

The eligibility criteria for Mortgage Pfandbriefe are specified in § 12 to § 18 PfandBG. Both commercial and residential mortgages can be included in a mortgage cover pool—as long as they fulfil certain eligibility criteria, such as:

- According to § 13 par. 1 PfandBG, the mortgages "must encumber properties, rights equivalent to real property or such rights under a foreign legal system that are comparable with rights equivalent to real property under German law". Furthermore, they must be located in the European Union (EU), in the European Economic Area (EEA), or in Switzerland, the US, Canada or Japan. For mortgage loans from outside the EU where the preferential claim of Pfandbrief holders in case of issuer insolvency is not ensured, a limit of 10 % applies.
- Mortgages can only be included in the cover pool up to 60 % of the so-called mortgage lending value (§ 14 PfandBG).[2] A loan exceeding the 60 % limit can be included, but only up to this limit. The part exceeding the limit is not eligible for cover and cannot be accounted for in the matching cover calculations as specified in Sect. 2.8 (Deutscher Bundestag 2008, p. 31). For more details on the mortgage lending value, see Sect. 2.7.
- The property underlying the mortgage must be insured against risks relevant to the location and type of the property (§ 15 PfandBG) such as, for example, fire.
- According to § 16 PfandBG, mortgages on building sites and new buildings under construction (which are not yet capable of producing an income) are eligible, too, but both of them together are not allowed to exceed 10 % of cover pool assets (or twice the Pfandbrief bank's liable own capital, whichever is higher). Mortgages on building land are not allowed to exceed 1 % of cover pool assets. Furthermore, mortgages on properties that do not produce income on an ongoing basis (such as pits and quarries) are not allowed into the cover pool at all. The same applies to mortgages on mines.
- Assets held in a fiduciary capacity ("treuhänderischer Verwaltung") by another credit institution are also eligible, as long as the Pfandbrief bank is entitled to separate these assets in the event of the trustee's insolvency (§ 1 par. 2 PfandBG). In this context, the refinancing register (see Sect. 2.6) becomes relevant.

Public Pfandbriefe

According to § 20 of the German Pfandbrief Act, public sector cover pools are allowed to contain claims (i.e. bonds, loans or comparable legal transactions) against public sector debtors, which according to Tolckmitt and Stöcker (2012b) include

- EU and EEA member states, including their regional governments and local authorities,

---

[2] For more details on the mortgage lending value, see Sect. 2.7.

- Switzerland, Japan, Canada and the US, including their regional governments and local authorities (subject to the requirement that they qualify for *credit quality step* 1),
- public sector entities of an EU/EEA member state,
- certain public sector entities from Switzerland, Japan, Canada and the US (subject to the requirement that they qualify for credit quality step 1),
- German public corporations and public-law institutions for which state support *(Anstaltslast)*, a legally founded guarantee obligation *(Gewährträgerhaftung)* or a state refinancing guarantee exists, or which are authorized to impose fees, rates and other levies.
- the European Central Bank, multilateral development banks and certain international organizations in terms of the EU Banking Directive (subject to the requirement that they qualify for credit quality step 1),
- central banks of the aforementioned states (for Switzerland, Japan, Canada and the US subject to the requirement that they qualify for credit quality step 1), and
- export credit agencies fulfilling certain additional requirements.

The claims must be free from any pleas. As for Mortgage Pfandbriefe, the Pfandbrief Act sets a limit of 10 % for debt from outside the EU where the priority claim of the Pfandbrief holder is not ensured. For claims against Switzerland, Japan, Canada and the US and other debtors that are subject to the credit quality step 1 requirement, a rating downgrade does not immediately trigger ineligibility as long as the claim still qualifies for credit quality step 2. The total amount of assets qualifying for credit quality step 2 (but not for credit quality step 1), however, is not allowed to exceed 20 % of all outstanding Public Pfandbriefe.

Further Cover Assets

To increase cover pool liquidity without changing the basic characteristics of the respective cover pools, § 19 and § 20 PfandBG allow—to a certain extent—the inclusion of *further cover assets*:

- Claims against suitable credit institutions: Both public sector and mortgage cover pools can contain claims against suitable credit institutions[3] up to a limit of 10 % of the respective outstanding Pfandbriefe. The amount of the claims of the Pfandbrief bank must, however, be known at the time of purchase, and the exposure to a single credit institution is limited to 2 % of the respective outstanding Pfandbriefe.

Mortgage cover pools can also contain the following assets:

- Certain liquid assets: Besides claims against suitable credit institutions as mentioned above, certain other liquid assets that qualify as excess cover can be

[3] Such as the European Central Bank, a central bank of the member states of the European Union, or credit institutions as specified in § 4 par. 1 sent. 2 no. 3 PfandBG.

included,[4] too, but there is a total limit of 10 % for all these claims (including claims against suitable credit institutions).

- Public sector bonds: Certain public sector bonds that fulfil the eligibility criteria given in § 20 par. 1 PfandBG can also be included, but the total amount of these claims plus the liquid assets from the bullet point above may not exceed 20 % of the outstanding Mortgage Pfandbriefe.

Derivatives

According to § 19 par. 1 no. 4 and § 20 par. 2 no. 3, derivatives can be included in mortgage and public sector cover pools. Derivatives in the sense of the Pfandbrief Act are derivatives under a standardized master agreement according to § 1 par. 11 sent. 4 no. 1 KWG, including the collateral support annexes concluded under the master contract and further agreements (§ 4 par. 3 sent. 2 PfandBG). To be cover pool eligible, they must fulfil the following additional requirements:

- Cover pool derivatives may only contain or replicate risks that are also inherent in eligible cover assets; open option short positions and positions with similar risk profiles are explicitly forbidden. The literature mostly mentions interest rate and currency swaps in the context of cover pool derivatives (see, for example, Tolckmitt and Stöcker 2012b). According to Volk (2011a), p. 77, credit derivatives are not allowed with the only exception being some kinds of credit linked notes under certain conditions.
- The derivative transaction must be concluded with "suitable credit institutions, investment management companies, public investment companies, financial services institutions, insurance companies, a central counterparty at a stock exchange, the German Federal Government or the Federal States" (§ 19 par. 1 no. 4).
- It must be ensured that the derivative does not contain any termination clause that applies in case of the Pfandbrief bank's insolvency. Tolckmitt and Stöcker (2012b) explain how this is achieved in practice: Individual agreements are kept under one master agreement—one for each cover pool, and one for the rest of the assets on the issuer's balance sheet. Netting then takes place only within each individual agreement. In case of issuer insolvency, the individual agreement for the other assets on the issuer's balance sheet is terminated, while the other two individual agreements remain in place and continue until the respective cover pool becomes insolvent. Collateral provided by the derivative counterpart (or by the Pfandbrief bank, if applicable) also needs to be registered in the cover register to ensure that it belongs to the separate legal estate in case of issuer insolvency (§ 13 DeckRegV).
- On a net present value basis, the Pfandbrief bank's claims from derivatives may not exceed 12 % of the claims against the cover pool, and the Pfandbrief bank's liabilities from derivatives are limited to 12 % of the sum of outstanding Pfandbriefe and liabilities arising from derivatives. The 12 % limit does not apply to derivatives that are solely used to hedge currency risks.

[4] Subject to the requirement that they fulfil § 4 par. 1 sent. 2 nos. 1 and 2 PfandBG.

In case of issuer insolvency, potential claims of derivative counterparties rank pari passu with the Pfandbrief holder's claims against the cover pool, and—in case of cover pool insolvency—against the issuer's general insolvency estate. For more details, see Sect. 2.12.

2.6 Refinancing Register

The *refinancing register* was introduced into the German Banking Act in 2005, and one of its purposes is to facilitate refinancing through Pfandbrief issuance and securitization. As already mentioned in Sect. 2.5, the PfandBG allows Pfandbrief banks to use mortgage loans held in a fiduciary capacity by a third party as cover assets for their own Mortgage Pfandbriefe if it can be ensured that, in case of the other credit institution's insolvency, these assets can be segregated from the insolvency estate. This can be achieved by entering the corresponding assets in a refinancing register. With this arrangement, Pfandbrief banks can use (portfolios of) mortgage loans originated and administered by other credit institutions for their Mortgage Pfandbriefe without the need to transfer the mortgages first, which is typically both costly and time-consuming (S&P 2007b). Details regarding the administration and the maintenance of the refinancing register are regulated in the RefiRegV. For further explanations on the role of the refinancing register in the context of Pfandbriefe see, for example, DG Hyp (2012), Fitch (2010a), or S&P (2007b).

2.7 Mortgage Lending Value

According to § 3 BelWertV, the *mortgage lending value*, which is addressed by § 14 and § 16 PfandBG, is defined as follows:

> (1) The value on which the lending is based (mortgage lending value) is the value of the property, which based on experience, may throughout the life of the lending be expected to be generated in the event of sale, unattached by temporary, e.g. economically induced, fluctuations in value on the relevant property market and excluding speculative elements.
> (2) To determine the mortgage lending value, the future marketability of the property is to be taken as a basis within the scope of a prudent valuation, by taking into account long-term sustainable aspects of the property, the normal and local market conditions, the current use and alternative appropriate uses of the property.

Detailed requirements regarding principle and procedure of its determination (i.e. the property valuation process itself and the methods to be used), the qualifications and the independence of the valuer, the mortgage lending value report[5] and re-evaluation triggers are specified by the BelWertV. While the market value of a real

[5] The mortgage lending value report reflects the valuer's opinion of the property value and is the basis for the determination of the mortgage lending value by a credit officer (Crimmann and Rüchardt 2008, p. 81).

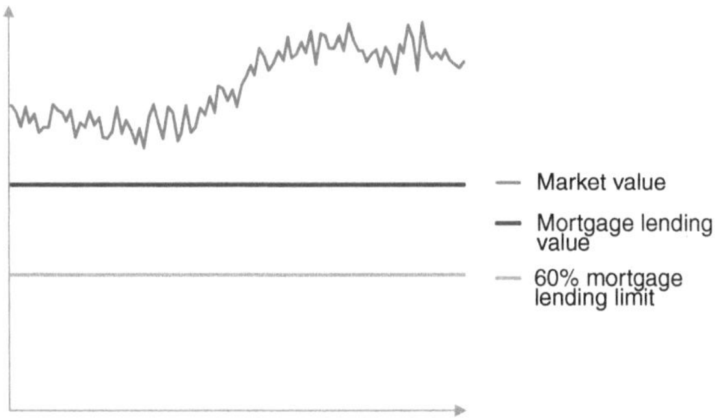

Fig. 3 Mortgage lending value versus market value (see Tolckmitt and Stöcker 2012b)

estate property is related to a specific valuation date (and, therefore, volatile and dependent on the business-cycle), the mortgage lending value refers to the whole life of the lending. It is constructed such that it does not fall below the market value. In times of booming markets, it should differ more from the market value than in times of economic recessions (Crimmann and Rüchardt 2008, p. 88). Figure 3 illustrates the characteristics of the mortgage lending value in the context of the 60 % mortgage lending limit specified in Sect. 2.5.

According to § 26 BelWertV, the mortgage lending value needs to be reviewed when there are indications that the market environment has deteriorated significantly. It is also specified when this applies:

> [T]his applies when the general price level in the respective regional property market has fallen to an extent that jeopardizes the safety of the lending. Unless owner-occupied residential properties are concerned, a review must also be conducted if the claim secured by the property serving as collateral shows substantial payment arrears of at least 90 days.

In practice, this means that a revaluation is necessary in case of market losses of more than 10 % (20 %) for commercial (residential) real estate properties over a period of up to three years (Dresch 2010). Payment arrears can be considered to be substantial when they account for 10 % of the outstanding loan amount (Koppmann 2009, p. 183). There are also further review requirements resulting from § 20a par. 6 KWG[6]:

- Under normal circumstances, commercial real estate needs to be re-evaluated at least once a year, while residential real estate has to be re-evaluated at least once every three years. In case the corresponding property market is very volatile, the re-evaluation has to be carried out more often. Institutions are allowed to use statistical methods to monitor property values and to identify re-evaluation needs.
- A real estate property needs to be re-valued by an independent valuer when there is evidence that its value has decreased considerably as compared to the general market value of comparable real estate.

[6] § 20a par. 6 KWG contains provisions regarding the revaluation of real estate collateral in the context of covered bonds. (According to § 20a, par. 1, sent. 1, no. 1, Pfandbriefe are a form of covered bonds.)

- In case the mortgage exceeds EUR 3 mn or 5 % of the credit institution's equity, a re-evaluation by an independent valuer is required at least every three years.

If necessary, the mortgage lending value has to be adjusted, which leads to a reduced recognition of the corresponding asset in matching cover calculations (see Sect. 2.8).

The financial crisis has proved that the mortgage lending value is, indeed, very stable: "Despite the falling prices in many property markets, there was no need to constantly revalue the loans, which was not the case had the market value procedure been used" (Baumgartner 2012). For more information on the mortgage lending value see, for example, Crimmann and Rüchardt (2008), Baumgartner (2012), Quentin (2008).

2.8 Cover Requirements

§ 4 PfandBG specifies several cover requirements that need to be fulfilled at any time and separately for each Pfandbrief type. These requirements are explained in more detail in the following.

- *Nominal cover:* The nominal amount of all outstanding Pfandbriefe must be covered by a cover pool of at least the same nominal amount. When the maximum redemption value of a Pfandbrief is higher than the current redemption value, the maximum redemption value needs to be covered.
- *Excess cover:* The net present value of all outstanding Pfandbriefe (including interest and amortization commitments) must be covered at any time.[7] Furthermore, the net present value of the cover pool must exceed the net present value of the liabilities by at least 2 %. This excess cover must consist of highly liquid assets as specified by § 4 par. 1 PfandBG, and it must be sufficient to withstand certain interest rate and currency stress scenarios as specified by the PfandBarwertV.[8] In case the net present value cover is not met under the stress scenario, additional collateral needs to be posted. While net present value coverage has to be monitored daily, coverage under the stress scenarios must be checked on a weekly basis only. The limits regarding further cover assets as specified in Sect. 2.5 do not apply to the excess cover.

[7] The calculation of the net present value is specified in detail in the PfandBarwertV: It is obtained by discounting all future cash flows with the currency-specific yield curve for swap deals. Foreign exchange positions are then converted to euros by using the appropriate exchange rates. For derivatives, market values are used.

[8] Basically, there are three potential methods to determine the interest rate and currency stresses. They all shift the interest rate curve up and down by a certain amount of basis points, set resulting negative interest rates to zero, and apply a stressed exchange rate for positions in foreign currencies. The Pfandbrief bank can choose between a static approach, a dynamic approach, and a method based on the bank's internal risk model. More details can be found in the vdp's legislative materials vdp (2011d).

- *Liquidity buffer:* Liquidity gaps within the next 180 days must be covered by liquid assets. For each day, the liquidity gap is calculated as the difference between scheduled payments (to Pfandbrief holders and cover pool derivative counterparties) and cash collections (from cover pool assets and cover pool derivatives). The maximum cumulative liquidity gap occurring within the next 180 days must be covered by assets that are allowed as excess cover and by ECB eligible assets. For assets that are exclusively used to ensure liquidity, the limits as described in Sect. 2.5 do not apply.

Not only liabilities resulting from Pfandbrief issuance, but also liabilities resulting from cover pool derivatives must be covered at any time. To ensure that the required cover is always given, the Pfandbrief bank performs daily *matching cover calculations*. As long as assets fulfil the eligibility criteria from Sect. 2.5, they can be fully considered in these calculations. A sole decline in market prices does not automatically trigger a reduced recognition. Delinquent cover pool assets also can be fully recognized as long as there has been no write-down (Koppmann 2009, p. 220).[9] The purpose of the cover requirements is to make sure that the privileged creditors' claims can be satisfied at any time by providing a certain buffer against costs and cover pool risks arising after issuer insolvency. This buffer is referred to as *mandatory overcollateralization*. In practice, many Pfandbrief issuers maintain even higher levels of overcollateralization (OC), which is referred to as *voluntary overcollateralization*. Figure 4 illustrates the different kinds of overcollateralization. As a result of the cover requirements, the cover pool needs active management: In case of repaying, defaulting, maturing or simply deregistered cover assets, suitable replacements might be necessary to ensure that the cover requirements are still fulfilled. Before further Pfandbriefe can be issued, additional assets may have to be posted.

2.9 Supervision

As illustrated by Fig. 5, the supervision of Pfandbrief banks is based on the PfandBG and the KWG. Pfandbrief banks are supervised by BaFin and monitored by a cover pool monitor and at least one deputy (§ 7 par. 1 PfandBG).

BaFin

BaFin conducts regular cover pool audits which—as a rule—shall take place every 2 years (§ 3 PfandBG).[10]

---

[9] As mentioned in Sect. 2.9, information on delinquent loans in the cover pool must, however, be disclosed regularly.

[10] These cover pool audits are either performed by BaFin itself, or BaFin can mandate other persons or institutions for this purpose.

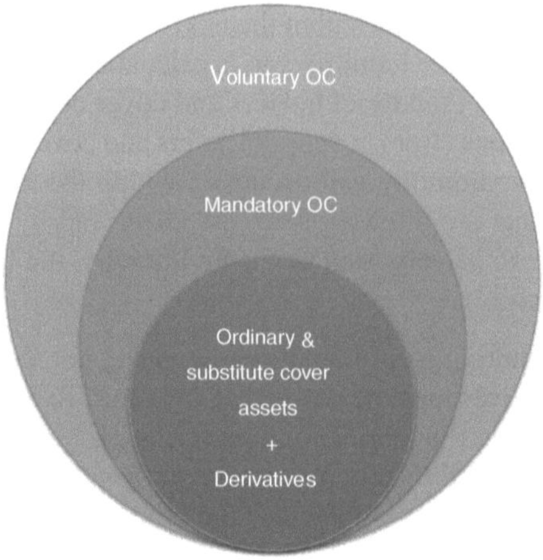

Fig. 4 The structure of overcollateralization

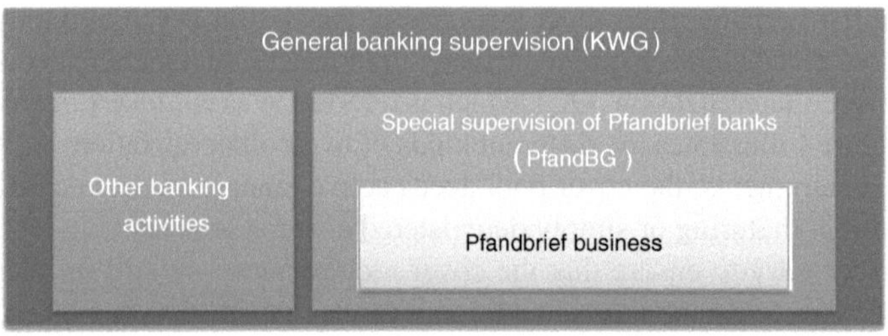

Fig. 5 Supervision of Pfandbrief banks (see Tolckmitt and Stöcker 2012b)

Cover Pool Monitor

In § 7 to § 11 PfandBG, detailed regulations regarding the appointment of a *cover pool monitor* ("Treuhänder"), his necessary qualification and his rights and responsibilities are set out. The cover pool monitor is independent, i.e. he is not bound by the instructions of the bank, BaFin or the Pfandbrief holders, and the sole objective of the tasks he performs is to safeguard the Pfandbrief's quality. According to § 8 PfandBG he has to ensure that

- the required cover exists at any time, especially when new Pfandbriefe are issued or when assets are deregistered,
- the cover register is complete and correct, and that derivative counterparties are informed about the registration, and that
- the mortgage lending value has been determined according to the Pfandbrief legislation.[11]

[11] He is, however, not obliged to check whether the fixed mortgage lending value is correct.

He needs to give his consent to each deregistration of a cover pool asset or derivative. As set out in § 7 PfandBG, he monitors the Pfandbrief bank's operations and reports his observations to BaFin. His function is discontinued when a cover pool administrator is appointed, but he then is still required to provide relevant knowledge to the cover pool administrator.

2.10 Risk Management Requirements

According to § 27 PfandBG, a Pfandbrief bank must have risk management systems in place that are appropriate to identify, assess, control and monitor the relevant risks arising from the Pfandbrief business, such as credit risks, interest rate risks, currency risks, other market risks, operational risks and liquidity risks. Before a Pfandbrief bank can engage in new markets, new products or new business activities, it has to perform a thorough risk analysis. The registration of assets related to these activities is only possible after sufficient knowledge and experience has been obtained, which is—in case of mortgage lending activities in new markets—at the earliest two years after taking up these activities.

2.11 Transparency

The *§ 28 transparency requirements* of the PfandBG specify key figures related to the Pfandbrief business that have to be published on a quarterly basis. These figures include:

- the total amount of outstanding Pfandbriefe and the related cover pools, given in terms of nominal value, net present value and stressed net present value,[12]
- the maturity profile of outstanding covered bonds and fixed-interest periods of cover assets, using pre-defined maturity ranges,
- the share of cover pool derivatives,
- the share of further cover assets (in accordance with § 19 par. 1 nos. 2 and 3 and § 20 par. 2 no. 2), and
- the amount of payments in arrears for at least 90 days, including their regional distribution.

Furthermore, there are additional Pfandbrief type specific reporting requirements addressing the granularity and the composition of the cover pool, such as

- type and size of the mortgage (for Mortgage Pfandbriefe),
- region and use of the property (also for Mortgage Pfandbriefe), and
- country and type of debtor/guarantor (for Public Pfandbriefe).

[12] The calculation of the net present value and the stressed net present value is defined in the PfandBarwertV, and explained in more detail in Sect. 2.8.

The reports must also include the corresponding previous year's figures. Once a year, additional information regarding foreclosures, receivership proceedings, property takeovers to prevent losses and arrears of interest payable not written off in the previous year has to be disclosed for Mortgage Pfandbriefe. The corresponding publications can also be found on the vdp's website (vdp 2013e).

The latest amendment of the PfandBG, which is scheduled to take effect on January 1st, 2014, contains further transparency requirements which include (cf. Bundesrat 2013, pp. 73–74)

- additional information on interest rate and currency mismatches,
- a more granular disclosure of Pfandbrief maturities and fixed interest rate periods for the first two years,
- further details on mortgage loans (the weighted-average loan-to-mortgage-lending value of the mortgage loan parts which have been recorded in the cover register, the weighted-average loan seasoning, and a more granular split of mortgage claim sizes),
- more information on the composition of further cover pool assets, and
- the disclosure of the amount of cover pool assets which are, in principle, cover pool eligible but exceed certain limits imposed by the PfandBG.

According to the transitional provisions of the amended PfandBG, the new transparency requirements have to be applied starting from Q2 2014 reporting (Bundesrat 2013, p. 77).

2.12 Issuer Insolvency

To ensure the preferential claim of the privileged creditors, § 30 to § 36a PfandBG specify in detail what happens in case of issuer insolvency. Until today, the working of these mechanics has, however, not yet been tested in practice. This is due to the reason that, in the past, Pfandbrief issuers that were close to insolvency were always bailed out (cf. Sect. 4.1.1).

Asset Segregation

According to § 30 PfandBG, upon issuer insolvency the cover pools are isolated from the issuer's general insolvency estate and do not take part in the bank's general insolvency proceedings (insolvency-free estate).[13] The segregation is based on the record in the cover register, and not on eligibility criteria applying at that point in time, i.e. assets recorded in the cover register that do not fulfil eligibility criteria are still part of the insolvency-free estate. As pointed out by Deutscher Bundestag (2008), p. 41, this ensures a clear separation of assets after issuer insolvency. The separation

[13] This implies that they are exempt from the German general insolvency law rule which states that insolvency proceedings concern all assets belonging to the corresponding debtor (Kullmann 2011).

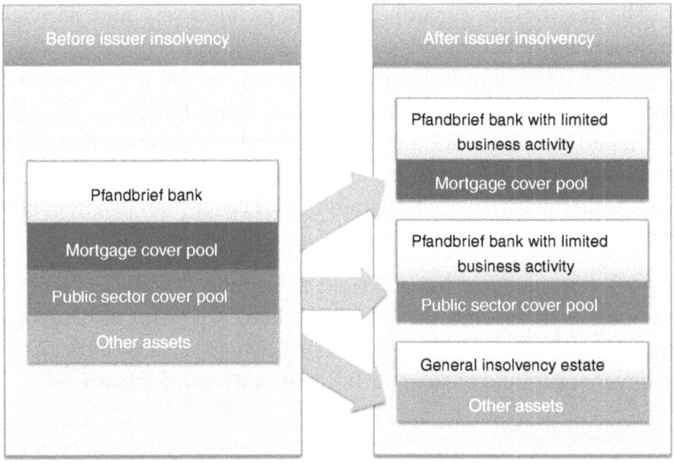

Fig. 6 Cover pool separation upon issuer insolvency

of the cover pool from the issuer's general insolvency estate is also referred to as *ring-fencing*.

In contrast to senior unsecured debt, Pfandbriefe do not automatically accelerate upon issuer insolvency.[14] The ring-fenced cover pool and the corresponding outstanding Pfandbriefe continue as a Pfandbrief bank with limited business activity ("Pfandbriefbank mit beschränkter Geschäftstätigkeit"), the purpose of which is the full and timely repayment of the privileged creditors. In case a Pfandbrief bank has issued more than one Pfandbrief type, there would be more than one Pfandbrief bank with limited business activity, each having a separate fate (Stöcker 2010). Figure 6 illustrates the separation principle.

Cover Pool Administrator

As set out in § 30 PfandBG, upon issuer insolvency—or even earlier if deemed necessary by BaFin—at least one *cover pool administrator* ("Sachwalter") is appointed. His main task is managing the cover pool and ensuring timely payments to the privileged creditors until their claims are fully satisfied. He is remunerated out of the cover pool. In his role as head of a Pfandbrief bank with limited business activity, the cover pool administrator can carry out the following legal transactions:

- He can use cover pool payments to service outstanding Pfandbriefe according to their contractual terms. He can also raise liquidity by refinancing activities such as central bank funding, take up refinancing loans or issue new bonds/Pfandbriefe. Furthermore, he has the possibility to raise funds by selling cover assets which is facilitated by the fact that he can use an existing or newly created refinancing

[14] This is another exemption from the German general insolvency law (Kullmann 2011).

register. The restrictions regarding further cover assets and derivatives from Sect. 2.5 do not apply to him.

- With the written consent of BaFin he can transfer the whole cover pool or parts of it as a package together with outstanding Pfandbriefe to another Pfandbrief bank. In case of a partial cover pool transfer, it needs to be ensured that the remaining part of the cover pool still fulfils the cover requirements.
- He can also hold the corresponding (part of the) cover pool in a fiduciary capacity for the other Pfandbrief bank.

The cover pool administrator has the right to use the Pfandbrief bank's staff and materials as long as he refunds the Pfandbrief bank's general insolvency estate for the occurring costs (§ 31 par. 8 PfandBG). Furthermore, he has to check on a regular basis if the cover requirements are still fulfilled (§ 31 par. 5 PfandBG).

Senior Unsecured Creditors

As specified by § 30 PfandBG, the cover pool is administered by the cover pool administrator, and the administrator of the issuer's general insolvency estate (called *insolvency administrator* in the following) has no access to it. Cash flows from the cover pool and cash flows from assets that only partially belong to the cover pool first go to the cover pool administrator, but the insolvency administrator can claim that these cash flows are separated at his expense. The insolvency administrator can also claim at any time that registered assets that are "obviously not [...] necessary" as Pfandbrief cover or mandatory overcollateralization are surrendered to him (§ 30 par. 4 sent. 1 PfandBG). Assets and derivatives remaining in the cover pool after all preferred creditors have been satisfied and after all administration costs (i.e. fees to the cover pool administrator and his expenses) have been deducted must be released to the issuer's general insolvency estate (§ 30 par. 4 sent. 2 PfandBG). Figure 7 summarizes the essential aspects of cover pool separation in case of issuer insolvency.

Fig. 7 Essential aspects of cover pool separation

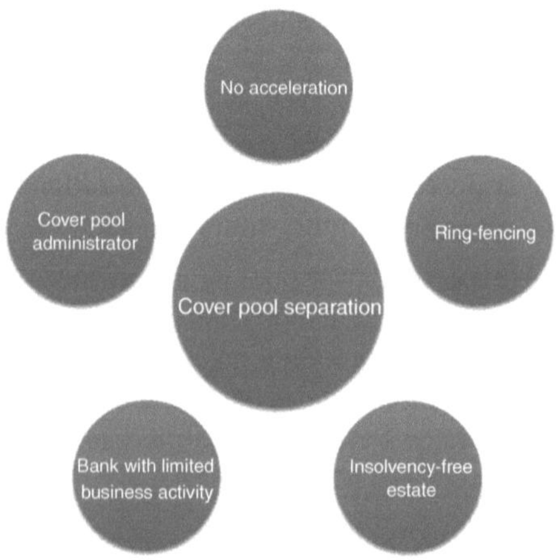

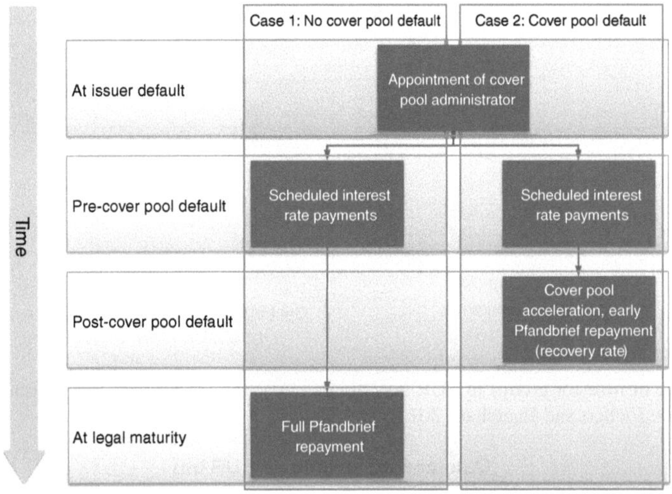

Fig. 8 Situation post-issuer insolvency

Cover Pool Insolvency

In case of *cover pool insolvency* (i.e. default or over-indebtedness), BaFin can initiate an insolvency proceeding for the cover pool. A cover pool insolvency administrator is then appointed, and the cover pool is accelerated. When the proceeds from cover pool liquidation are not sufficient to satisfy the preferred creditor's claims, they still have a claim against the issuer's general insolvency estate, i.e. they can participate in the issuer's general insolvency proceedings and rank pari passu with senior unsecured creditors. This is also specified in § 30 PfandBG. Figure 8 gives an overview of the situation post-issuer insolvency.

3 The Pfandbrief Market

This section gives a brief overview of the covered bond market and the role the German Pfandbrief plays in this market. It is mainly based on 2011 data published by the European Covered Bond Council (*ECBC*), which covers 26 jurisdictions (see ECBC 2012). Unless otherwise stated, we refer to these numbers. For the German market, this data is based on Deutsche Bundesbank statistics and consistent with vdp data (see vdp 2012e).

The Covered Bond Market

With an outstanding volume of EUR 2,676 bn at the end of 2011 and a new issuance volume of EUR 695 bn in that year, covered bonds play an important role in European

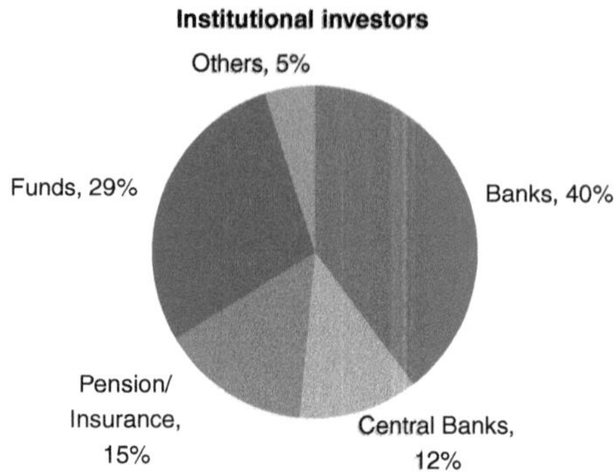

Fig. 9 Share of investor groups in EUR benchmark covered bond primary market transactions in 2011 (*Source*: Eichert and Engelhard 2012)

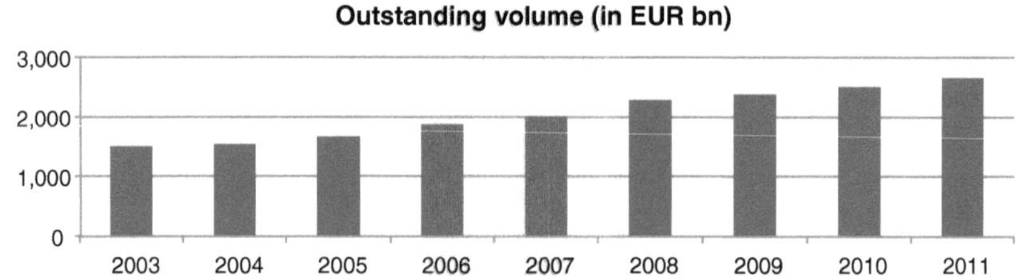

Outstanding covered bond volume (in EUR bn) and number of issuers									
Year	**2003**	**2004**	**2005**	**2006**	**2007**	**2008**	**2009**	**2010**	**2011**
Outstanding volume	1.498	1.553	1.676	1.881	2.022	2.279	2.394	2.503	2.676
Number of issuers	140	167	194	213	233	266	299	301	319

Fig. 10 Outstanding covered bond volume and number of issuers over time

capital markets. The covered bond market's most liquid sub-segment, the *benchmark covered bond* market, constitutes one of the largest segments of the European bond market, second to government bonds (Grossmann and Stöcker 2012).

Covered bonds are mainly bought by institutional investors such as banks, central banks, pension funds, insurance companies and investment funds. Figure 9 shows the relative share of these investor groups in benchmark covered bond primary market transactions in 2011.[15] MEAG (the asset manager of MunichRe and Ergo), for example, which is one of the most important asset management companies in Europe, had roughly EUR 220 bn of assets under management in 2012, out of which EUR 55 bn, i.e. 25 %, were covered bonds. With EUR 30 bn, more than half of these covered bonds were Pfandbriefe (Wolf 2012).

The covered bond market has been growing continuously over the last years. As shown by Fig. 10, the outstanding volume of covered bonds has increased from EUR

[15] Retail investors fall into the banks' share.

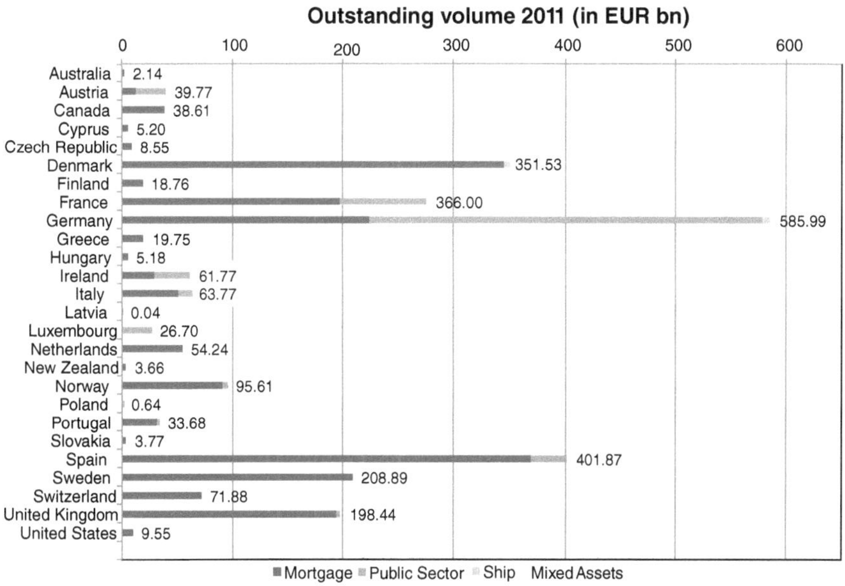

Fig. 11 Volume of outstanding covered bonds in 2011, by country and split into different collateral types

1,498 bn in 2003 to EUR 2,676 bn in 2011 (+79 %). The number of issuers has also risen, from 140 to 319. Nowadays, covered bonds are issued in almost all European countries, and also in non-European countries like Australia, Canada, New Zealand and the United States. From Fig. 11 it can be seen that, at the end of 2011, the German Pfandbrief market was the largest covered bond market with an outstanding volume of EUR 586 bn, followed by Spain (EUR 402 bn), France (EUR 366 bn) and Denmark (EUR 352 bn). The market is dominated by issuers from Germany, Spain, France and the Nordics,[16] which in 2011 accounted for 76 % in terms of outstanding volume, while issuers from the rest of Europe had a share of 22 %. Non-European issuers only played a minor role (2 %). Figure 12 shows the share of these different country groups in outstanding volume and new issuance in 2011.[17] The overall country composition of the covered bond market is, however, likely to change in future. Newer covered bond markets such as Australia, Canada and New Zealand have started growing (Eichert 2012) and, according to the statistics in ECBC (2012), the Italian, the Finish and the French market have also shown considerable growth in recent years. In addition to that, new issuers are likely to join the market. The number of countries with a covered bond legislation has increased from 29 in 2007 to 37 in 2012 (with two recent newcomers being Belgium and Chile), and there are ongoing legislative procedures in the US, Canada and South Korea (Immobilien

[16] By Nordics, we refer to Denmark, Sweden, Norway and Finland.

[17] Note that Danish issuers often roll over short dated bullet covered bonds, which leads to high volumes of new issuance in this country which, over time, do not necessarily result in increasing covered bond volumes (Eichert 2012).

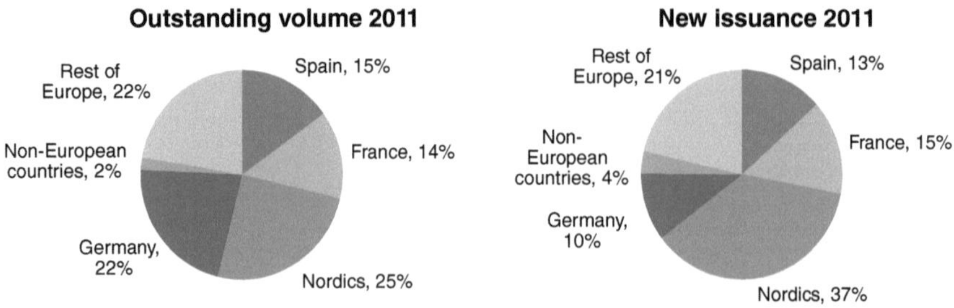

Fig. 12 Share of outstanding covered bond volume and new issuance in 2011, split by country groups

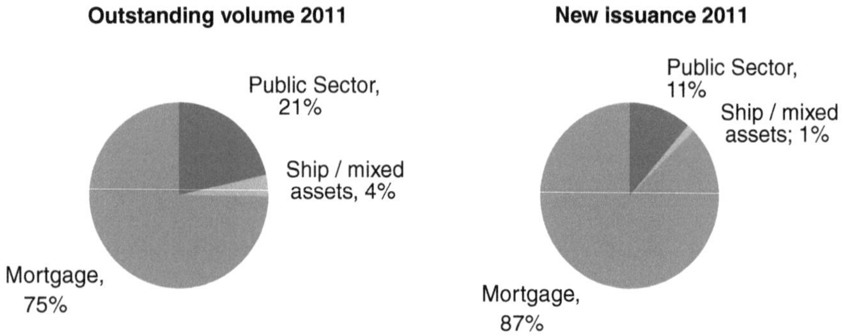

Fig. 13 Share of outstanding covered bond volume and new issuance in 2011, split by collateral type

& Finanzierung 2013). Countries like Brazil, Singapore and Morocco also plan to introduce covered bond laws. The German market, on the other hand, continues to shrink (see discussions below).

Figure 11 also shows that covered bonds are mainly used to refinance mortgage loans. All 26 jurisdictions except for Luxembourg have mortgage covered bonds outstanding.[18] Public sector covered bonds, in contrast, only exists in less than half of the considered countries, and covered bonds backed by mortgages on ships and mixed assets are very rare.[19] This is also depicted in Fig. 13 which shows that, in 2011, three quarters of outstanding covered bonds were backed by mortgage assets, while public sector collateral accounted for only 21 %. The aggregated share of mortgages on ships and mixed assets was 4 %. In new issuances, mortgage covered bonds had a share of 87 %, while 11 % were public sector covered bonds, and the remaining 1 % either backed by mortgages on ships or mixed assets.

[18] In Luxembourg, covered bonds backed by mortgage collateral ("Lettres de Gage Hypothécaires") do—in theory—exist, but according to the statistics, their outstanding volume has been zero since 2009.

[19] While covered bonds backed by mortgages on ships exist only in Germany and in Denmark, cover pools with mixed assets (meaning that there is more than one type of collateral) can only be found in France.

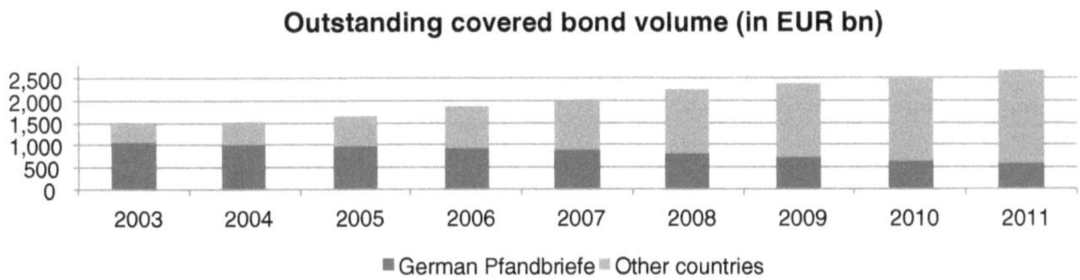

Outstanding covered bond volume (in EUR bn)									
Year	**2003**	**2004**	**2005**	**2006**	**2007**	**2008**	**2009**	**2010**	**2011**
German Pfandbriefe	1,057	1,010	976	949	889	806	719	640	586
Other countries	442	543	700	932	1,133	1,473	1,675	1,863	2,090
Total	1,498	1,553	1,676	1,881	2,022	2,279	2,394	2,503	2,676
% German Pfandbriefe	71%	65%	58%	50%	44%	35%	30%	26%	22%
% Other countries	29%	35%	42%	50%	56%	65%	70%	74%	78%

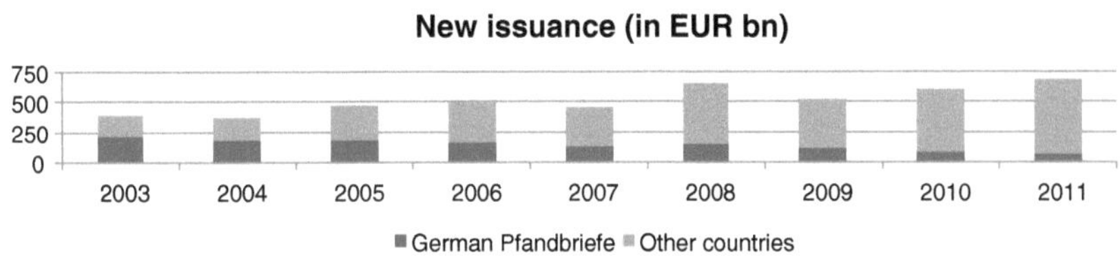

New issuance (in EUR bn)									
Year	**2003**	**2004**	**2005**	**2006**	**2007**	**2008**	**2009**	**2010**	**2011**
German Pfandbriefe	211	174	173	167	135	153	101	87	73
Other countries	188	199	297	331	329	497	421	526	623
Total	400	373	469	498	464	650	531	613	695
% German Pfandbriefe	53%	47%	37%	34%	29%	24%	21%	14%	10%
% Other countries	47%	53%	63%	66%	71%	76%	79%	86%	90%

Fig. 14 Share of German Pfandbriefe in total covered bond volume outstanding and new issuance

As already mentioned before, in 2011 the Pfandbrief market was the largest covered bond market. However, as can be seen from Fig. 14, the share of Pfandbriefe in total covered bonds has constantly decreased since 2003, both in terms of outstanding volume (where the share has gone down from 71 to 22 %), and in terms of new issuance (where the share has fallen from 53 to 10 %). There are two main reasons for this trend, the first one being that more and more covered bonds from other countries have been issued. While, in 2003, only EUR 442 bn of covered bonds from other countries were outstanding, this number has risen to EUR 2,090 bn in 2011 (+373 %). Over the same period, new issuances from these countries have increased from EUR 188 bn to EUR 623 bn (+231 %). In addition to that, the outstanding Pfandbrief volume and new issuances from Germany have decreased over the years (−45 and −66 % respectively). The reasons for this trend will be discussed in more detail in the following.

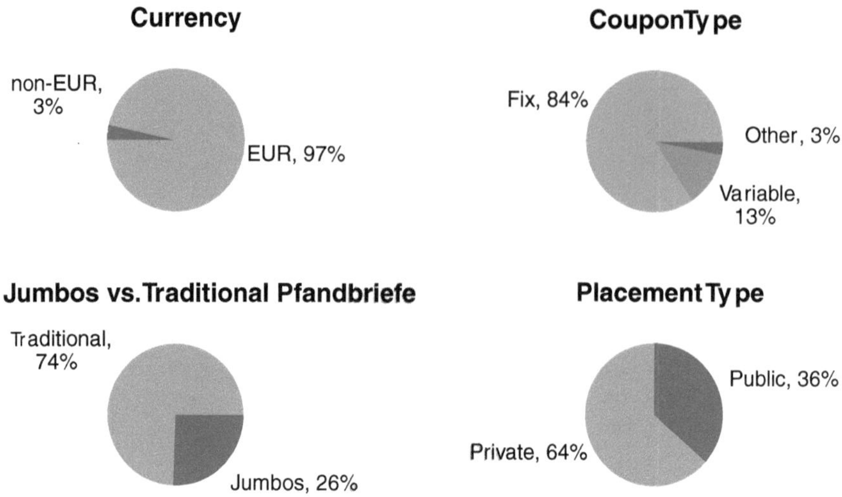

Fig. 15 Issuance type and profile of outstanding Pfandbriefe in 2011

The German Pfandbrief Market

In 2011, the German Pfandbrief market was the largest covered bond market with an outstanding volume of EUR 586 bn, and a new issuance volume of EUR 73 bn. As can be seen from Fig. 15, almost all of these outstanding Pfandbriefe (97 %) were denominated in euros, and most of them (84 %) carried a fix coupon. With 74 %, the majority was brought to the market in the form of *traditional Pfandbriefe*, while the remaining 26 % were issued as Jumbos. There were also more private placements (64 %) than public placements (36 %).

As detailed by Fig. 16, the outstanding Pfandbrief volume has dropped considerably over the last years, from EUR 1,057 bn in 2003 to EUR 586 bn in 2011 (−45 %), and new issuance has fallen from EUR 211 bn to EUR 73 bn (−66 %).[20] While Mortgage Pfandbriefe have remained comparably stable over time, the outstanding volume of the Public Pfandbrief segment has gone down from EUR 797 bn to EUR 356 bn (−55 %). Even though the share of outstanding Pfandbriefe backed by public sector collateral was still 61 % in 2011, the segment has lost ground since 2003, when its share was 76 %. Its share in new issuances used to be much higher, too (72 % in 2003, as compared to 43 % in 2011). The stable development of the Mortgage Pfandbrief segment can be attributed to the fact that Mortgage Pfandbriefe offer attractive refinancing possibilities. Cheap funding can be obtained at comparably low costs, and new issuers (saving banks and international players such as the German subsidiary of the French Natixis bank) are joining the market (Wolf 2012). The Public Pfandbrief segment, on the contrary, keeps declining, which is due to several reasons (see, for example, DG Hyp 2008; vdp 2010b; NordLB 2011). Public

[20] Note that the percentages in the table do not always sum up to 100 % due to rounding effects.

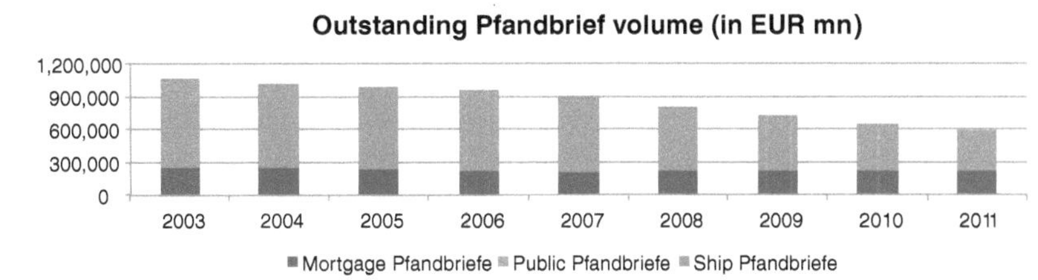

Outstanding Pfandbrief volume (in EUR mn)									
Year	**2003**	**2004**	**2005**	**2006**	**2007**	**2008**	**2009**	**2010**	**2011**
Mortgage Pfandbriefe	256,027	246,636	237,547	223,306	206,489	217,367	225,100	219,947	223,676
Public Pfandbriefe	797,492	760,264	734,713	720,835	677,656	578,974	486,406	412,090	355,673
Ship Pfandbriefe	3,172	3,212	3,670	4,669	4,413	9,282	7,954	7,805	6,641
Total	1,056,691	1,010,112	975,930	948,810	888,558	805,623	719,460	639,842	585,990
% Mortgage Pfandbriefe	24.2%	24.4%	24.3%	23.5%	23.2%	27.0%	31.3%	34.4%	38.2%
% Public Pfandbriefe	75.5%	75.3%	75.3%	76.0%	76.3%	71.9%	67.6%	64.4%	60.7%
% Ship Pfandbriefe	0.3%	0.3%	0.4%	0.5%	0.5%	1.2%	1.1%	1.2%	1.1%

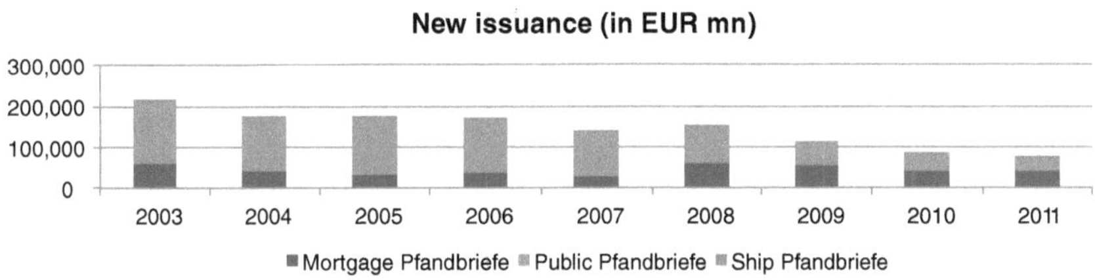

New issuance (in EUR mn)									
Year	**2003**	**2004**	**2005**	**2006**	**2007**	**2008**	**2009**	**2010**	**2011**
Mortgage Pfandbriefe	57,621	40,773	33,722	35,336	26,834	57,345	56,852	42,216	40,911
Public Pfandbriefe	151,690	131,506	137,235	129,452	107,913	89,522	52,251	41,574	30,990
Ship Pfandbriefe	2,103	1,646	1,742	2,374	628	6,054	1,286	3,189	895
Total	21,414	173,925	172,699	167,162	135,375	152,921	110,389	86,979	72,796
% Mortgage Pfandbriefe	27.3%	23.4%	19.5%	21.1%	19.8%	37.5%	51.5%	48.5%	56.2%
% Public Pfandbriefe	71.8%	75.6%	79.5%	77.4%	79.7%	58.5%	47.3%	47.8%	42.6%
% Ship Pfandbriefe	1.0%	0.9%	1.0%	1.4%	0.5%	4.0%	1.2%	3.7%	1.2%

Fig. 16 Share of different Pfandbrief types in oustanding volume and new issuance

sector lending has, in general, become less attractive. Decreasing profitability caused by low margins over the years and increasing regulatory requirements have caused issuers to reduce their activities in the public sector business, or to withdraw from it all together. The buyback programs of Berlin-Hannover Hypothekenbank (Berlin Hyp) and Dexia Kommunalbank in 2012 (cf. Wolf 2012) are recent examples of this trend. Furthermore, the implementation of the EU commission's requirements after the financial crisis has forced two of the largest issuers, Deutsche Pfandbriefbank and Hypothekenbank Frankfurt (former Eurohypo), to reduce their balance sheets and, in this context, their public sector business. The decline of the Public Pfandbrief

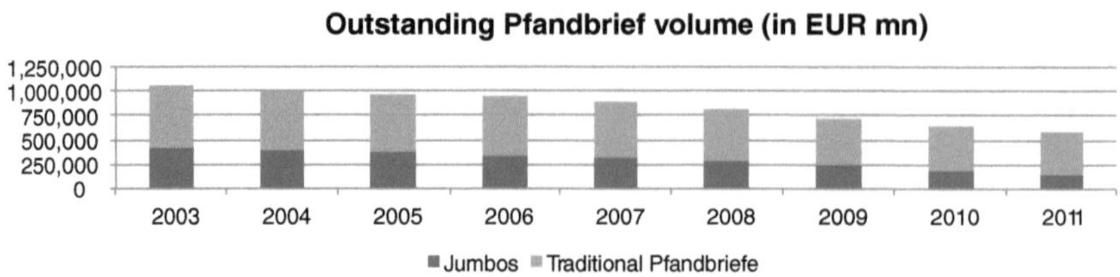

Outstanding Pfandbrief volume (in EUR mn)									
Year	2003	2004	2005	2006	2007	2008	2009	2010	2011
Jumbos	413,700	391,400	372,600	345,640	312,358	279,176	233,500	178,818	149,752
Traditional Pfandbriefe	642,991	618,712	603,330	603,170	576,200	526,447	485,960	461,024	436,238
Total	1,056,691	1,010,112	975,930	948,810	888,558	805,623	719,460	639,842	585,990
% Jumbos	39.2%	38.7%	38.2%	36.4%	35.2%	34.7%	32.5%	27.9%	25.6%
% Traditional Pfandbriefe	60.8%	61.3%	61.8%	63.6%	64.8%	65.3%	67.5%	72.1%	74.4%

Fig. 17 Share of Jumbos and traditional Pfandbriefe in outstanding Pfandbrief volume

segment is also impacted by the fact that there are generally less assets available to be included in public sector cover pools, caused by the abolition of state guarantees in the form of Gewährträgerhaftung in 2001, and the removal of debt from periphery countries (such as Portugal, Italy, Ireland, Spain and Greece) from cover pools in the light of the sovereign crisis.[21] Fitch (2012a) expects the trend to dormant and wind-down Public Pfandbrief programs to continue in 2013. All in all, despite the stable development of the Mortgage Pfandbrief, the overall volume of outstanding Pfandbriefe is expected to decline further, due to a further shrinking of the Public Pfandbrief segment (Schönfeld 2012).

A second trend that can be observed in the Pfandbrief segment is related to issuance size. As shown by Fig. 17, the amount and the share of outstanding *Jumbo Pfandbriefe* has continuously decreased over time. While, in 2003, EUR 414 bn of Jumbos were outstanding, it was only EUR 150 bn in 2011. Their relative share in the Pfandbrief market has also decreased, from 39 to 26 %. As pointed out by Volk (2012), this reflects a recent trend towards benchmark issues with a volume of EUR 500 mn to EUR 1 bn. In comparison with Jumbos, which are considered to be more liquid than smaller issuance sizes, benchmark issues are more favourable in the context of asset liability management and legal requirements regarding liquidity holdings (Wolf 2012). Widened index eligibility,[22] higher-execution risks in the context of deal placement, lower funding needs and new market practices are further reasons for a trend towards smaller issuance sizes (Caris 2012).

[21] Before the PfandBG was introduced in 2005, (unsecured) claims against savings banks and Landesbanks, which benefited from state guarantees in the form of Gewährträgerhaftung, were cover pool eligible and represented high shares of public sector cover pools. Since the abolition of these state guarantees, claims against such credit institutions are, in general, not cover pool eligible any more. (Source: Golin 2006, p. 139)

[22] For indices, a size of EUR 500 mn is nowadays considered to be the norm for inclusion. Markit, for example, lowered the eligibility criteria for its iBoxx index accordingly in December 2011. (Source: Caris 2012)

At the time of writing, 2012 Pfandbrief statistics were already published on the vdp's website (see vdp 2013j). For the covered bond market, however, 2012 data was not yet available, which is why we decided not to include the Pfandbrief data in the above statistics. All in all, a continuation of the overall trends just discussed can be observed, i.e. comparably stable Mortgage Pfandbrief developments (with a small overall decrease), an accelerated decline in Public Pfandbriefe and increasing importance of benchmark issues. One particular point to mention is, that for the first time, an Aircraft Pfandbrief was issued in 2012 (by NordLB). For more details on the 2012 developments in the Pfandbrief market, we refer to the vdp's annual report (vdp 2013b).

Spread Development

German Pfandbriefe have performed well during the financial crisis and the euro-zone government debt crisis, especially when compared to covered bonds from other jurisdictions. iBoxx Covered spreads, which represent the covered bond universe, have widened considerably since 2008, while the spreads of iBoxx Mortgage Pfandbriefe and iBoxx Public Pfandbriefe have remained comparably stable and much narrower, see Fig. 18a. German Pfandbriefe also suffered from the Lehman collapse in September 2008, but when the ECB announced its covered bond purchase program in 2009, spreads almost returned to pre-crisis levels, with comparably low spread widenings since then (DG Hyp 2012). As opposed to covered bond issuers from European periphery countries, Pfandbrief issuers did not experience problems with market access in the recent past (Volk 2012). The spread differences between the iBoxx Pfandbrief indices on the one hand and iBoxx Covered on the other hand are now at a historic high,[23] i.e. Pfandbriefe offer very favourable funding conditions as compared to other covered bonds. According to DG Hyp (2012) this is due to the Pfandbrief's loyal and broad investor base and the faith in the product, but also due to the limited Pfandbrief supply caused by reduced issuance activities.

Pfandbriefe still offer an attractive yield pickup versus German government bonds, see Fig. 18b. According to Wolf (2012), this yield pickup of covered bonds versus sovereign debt is currently bigger and less volatile than in other jurisdictions and, therefore, highly appreciated by investors.

Summary and Outlook

Covered bonds play an important role on European capital markets and in the context of financial stability,[24] and the number of countries with a covered bond legislation

---

[23] As pointed out by Volk (2012) it should, however, be noted that the index data is based on secondary spreads only, meaning that it might not completely reflect market reality, especially as in periphery countries (such as Spain) a considerable portion of covered bonds is nowadays retained and refinanced through the ECB.

[24] See, for example, European Central Bank (2011): "A smoothly functioning covered bond market is highly important in the context of financial stability."

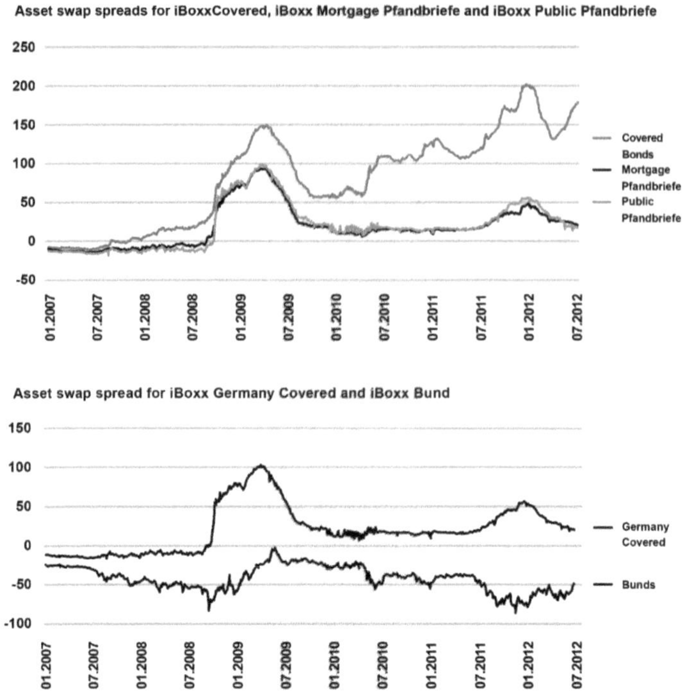

Fig. 18 Performance of Pfandbrief spreads in comparison to covered bonds and bunds (*Source*: Volk 2012). **a** iBoxx Mortgage/Public Pfandbrief versus iBoxx Covered, **b** iBoxx Germany covered versus iBoxx Bund

keeps growing. Within the covered bond market, the benchmark status of the German Pfandbrief is undisputed. During the financial crisis, the German Pfandbrief has proved to be a stable and reliable funding instrument. Even though its total outstanding volume is decreasing while issuance from other countries is on the rise, Mortgage Pfandbriefe are likely to play an important role in future (Bettink in vdp 2013d):

> The clear decline experienced by the Public Pfandbrief is the logical consequence of the regulation-driven withdrawal by many banks from public sector lending. This trend will continue. By contrast, 2012 was another good year on the whole for the Mortgage Pfandbrief, and its prospects for the future remain intact. The attractiveness of the product, for our member banks and for investors, is undiminished. This is borne out, not least, by the recent highly successful placements and the steadily growing interest shown by potential new issuers.

The number of Pfandbrief issuers, which currently is 72, has increased by almost 25 % over the last 4 years, and more issuers are expected to join the market in the near future (vdp 2013i). According to Wolf (2012), the importance of Pfandbriefe in institutional asset management has also grown in recent years.

Further Readings

For more information on recent developments in the Pfandbrief market, we suggest reading the vdp's annual *Pfandbrief Fact Book* (see vdp 2006b, 2007b, 2008c, 2009b, 2010d, 2011c, 2012d), and for a more general overview of the international covered bond market, the *European Covered Bond Fact Book* (see ECBC 2008, 2009, 2010, 2011, 2012) published by the European Covered Bond Council is recommended. Recent Pfandbrief market developments are also discussed by various research publications, such as Commerzbank's *Pfandbrief Weekly*.

4 Risk Analysis

As long as the Pfandbrief issuer is solvent, he is responsible for making full and timely Pfandbrief payments. Therefore, in the first place, the Pfandbrief holder is exposed to all kinds of risk related to the issuer (*issuer risks*). Upon issuer insolvency, the situation changes and cover pool performance becomes important as payments to the Pfandbrief holders are made out of the cover pool (*cover pool risks*). There are also risks associated with the timing of Pfandbrief redemption, structural and legal risks in the context of issuer and cover pool insolvency, and other risks that do not fit into the aforementioned categories. Figure 19 shows the relevance of these risks pre- and post-issuer insolvency. While, for obvious reasons, issuer risk only matters pre-issuer insolvency, the Pfandbrief holders become exposed to cover pool risks, the risk of timely Pfandbrief repayment and structural and legal risks after issuer insolvency only.

Figure 20 gives a detailed overview of the risks the Pfandbrief holder is exposed to. In the following, all these risks are analysed in detail. It is explained to what extent they are mitigated by the legal provisions of the PfandBG, through program-specific features such as voluntary overcollateralization and an issuer's commitment to his Pfandbrief program, or by further risk mitigants such as supervision, transparency, systemic support or flexibility of legislation. Our risk analysis is mainly based on the rating agency publications as listed in the References section, and uses an analysis by Moody's (2010d) as a starting point. A lot of helpful information on the topic can also be found in Volk (2009), Volk (2011a), Dübel (2010), and the more quantitative oriented publications from Siewert and Vonhoff (2011), Prokopczuk and Vonhoff (2012), Sünderhauf (2006) and Rudolf and Saunders (2009).

Type of risk	Pre-issuer insolvency	Post-issuer insolvency
Issuer risks	Yes	No
Cover pool risks	No	Yes
Risk of timely Pfandbrief repayment	No	Yes
Structural and legal risks	No	Yes
Other risks	Yes	Yes

Fig. 19 Relevance of Pfandbrief risks pre- and post-issuer insolvency

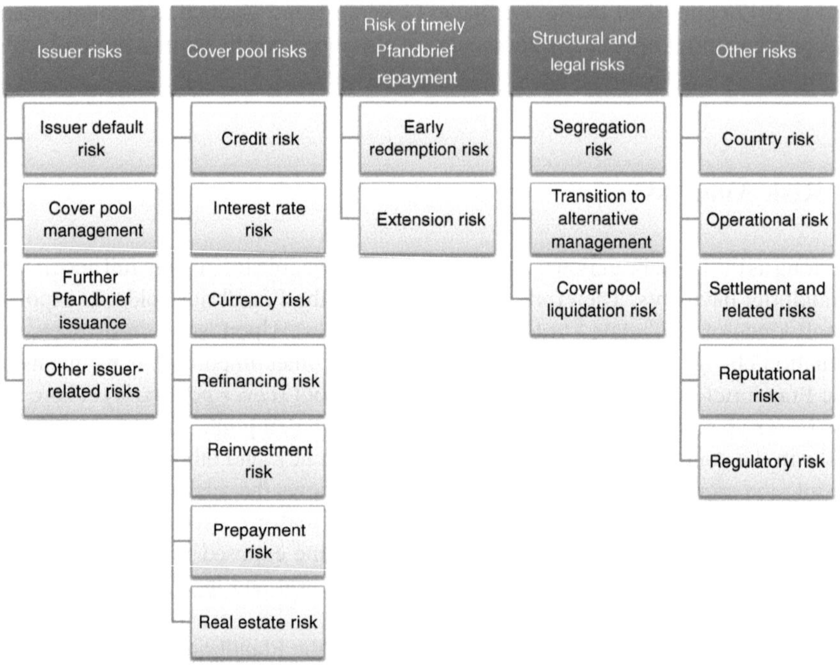

Fig. 20 Overview of Pfandbrief risks

4.1 Issuer Risks

The issuer is the primary source of the Pfandbrief holder's payments. Therefore, his ability to fulfil his payment obligations in a full and timely manner is essential. Furthermore, he has to make sure that cover requirements are fulfilled at any time and, if necessary, he needs to post additional collateral. The way he manages the cover pool is also important as the cover pool constitutes the Pfandbrief holder's secondary source of recourse. In addition to that, the issuer's decision regarding further Pfandbrief issuance, as well as other risks related to the Pfandbrief bank, may affect the Pfandbrief holder.

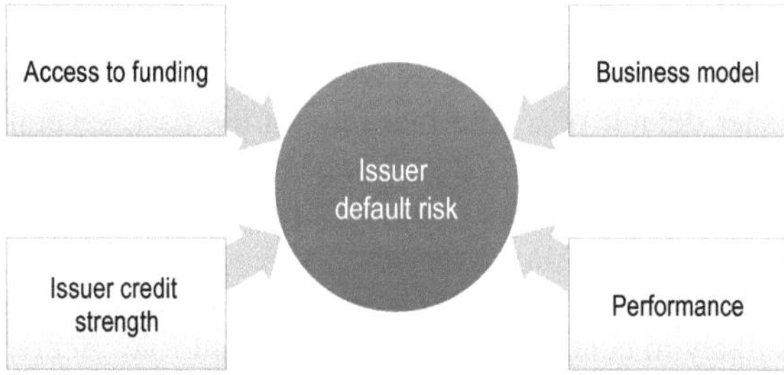

Fig. 21 Issuer default risk: sources of influence

4.1.1 Issuer Default Risk

The risk of losses due to the Pfandbrief issuer's failure to fulfil his payment obligations in a full and timely manner is referred to as *issuer default risk*. It is reflected in the Pfandbrief bank's senior unsecured rating. Fitch (2010), for example, considers the issuer's credit quality in general, his access to stable funding sources (*financial risk*), the business activities he is engaged in (*business risk*), and his performance regarding income stability, diversity and profitability (*performance risk*) when assigning a rating to the issuer, cf. Fig. 21. The higher the issuer's credit strength, the more likely it is that he is able to replace non-performing assets in the cover pool and to absorb losses. His credit strength also depends, to some extent, on the corresponding sovereign's credit strength, which is why bank ratings often account for a certain amount of sovereign support (see Sect. 4.5.1). The issuer's access to stable funding sources is especially important in the context of refinancing when there are large maturity mismatches in the cover pool: When credit lines to the issuer are cut in times of market distress, this might cause an issuer event of default when a large *bullet repayment* of a Pfandbrief cannot be refinanced. The issuer's business model and his general business activities also affect the Pfandbrief holder as they might threaten the solvency of the issuer. Under today's universal bank principle there are—apart from the (general) regulations of the German Banking Act—no restrictions regarding the business activities a Pfandbrief issuer is allowed to engage in.[25]

The importance of issuer default risk in the context of Pfandbriefe is shown by the fact that all three major rating agencies (i.e. Moody's, Fitch, and S&P) consider the role of the issuer and his rating when assigning Pfandbrief ratings (Melms 2012). According to Moody's (2010a), the issuer's creditworthiness plays a crucial role, and during the last years, issuer rating actions were often the driving force for covered bond rating actions (see Moody's 2008a, 2009c, 2010c, 2011a,b, 2012a). In recent years, the linkage between the ratings of issuers and covered bonds has, in general, become stronger (Melms 2012). In their paper, Packer et al (2007) find that covered bond prices are quite robust to issuer-specific credit risk shocks, but their analysis is based on pre-crisis data. A more recent study by Siewert and Vonhoff (2011), who use

[25] However, according to S&P (2005c), the former specialist bank principle did not have a deciding impact on Pfandbrief ratings.

regression analysis and distinguish three different time periods (pre-crisis, subprime crisis and post-Lehman), shows that investors require a considerable premium for issuer default risk. The fact that, in the past, several Pfandbrief issuer bailouts—such as in the case of Allgemeine Hypothekenbank Rheinboden AG (October 2005), Düsseldorfer Hypothekenbank (April 2008), Hypo Real Estate AG (October 2008), Eurohypo AG (May 2009) and Valovis Bank (end 2011)—were necessary (see Kiff et al. 2011; Die Welt 2012) clearly shows the relevance of issuer default risk, which is likely to continue to play an important role. According to Moody's (2011b), the credit quality of some German issuers has deteriorated in light of their exposures to weaker economies, and Moody's 2013 outlook for the German banking system is still negative (Moody's 2012a).[26]

4.1.2 Cover Pool Management

Cover pool management can have a big impact on the cover pool's quality, and the issuer's ability to change the cover pool's risk profile (which is referred to as *substitution risk*) and his decision regarding the maintenance of overcollateralization can affect the Pfandbrief holder in a negative way.

Substitution Risk

As long as eligibility criteria and cover requirements are fulfilled, the issuer can change the composition of the cover pool at any time by adding or removing cover pool assets, or by entering into new derivative contracts. This might have a negative impact on the cover pool's risk profile and increase cover pool risks. The credit quality of the cover pool could, for example, deteriorate due to asset replacements caused by the issuer's need to improve margins. Cover pool quality is, therefore, not independent from the issuer's solvency. The closer the issuer is to default, the more important substitution risk becomes. The issuer may generally have less high-quality assets or be forced to use good quality cover pool assets for other purposes and replace them by lower quality assets. In addition to that, he might not be willing or able any more to replace bad performing assets in the cover pool which still fulfil the eligibility criteria. In this context, the issuer's business model also plays an important role, as it defines the assets generally available for inclusion in cover.

The PfandBG mitigates substitution risk by eligibility criteria and monitoring requirements: only high quality assets are cover pool eligible, and the cover pool monitor has to give his consent to the deregistration of cover pool assets and derivatives. As pointed out by Moody's (2010d) this does, however, not extinguish substitution risk, as there is still a certain amount of freedom to swap cover pool assets. According to the rating agency (see Moody's 2012h), the risk of (adverse) asset selection

[26] Moody's banking system outlook reflects the rating agency's view of fundamental credit conditions which will impact the corresponding banks' creditworthiness over the next 12–18 months.

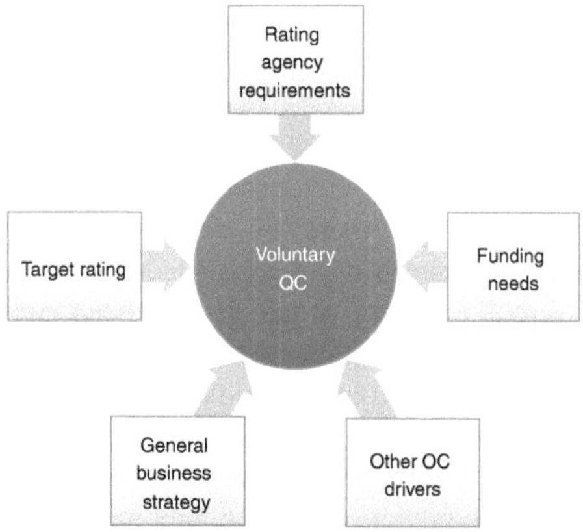

Fig. 22 Drivers of voluntary overcollateralization

has increased over the last years, especially in the case of Public Pfandbriefe which are allowed to contain assets from Greece or other periphery/sub-investment grade countries. To judge substitution risk appropriately, detailed information on (planned changes in) the business strategy, as well as on cover and non-cover assets, is needed. The availability of this information is, however, limited and often only available with a certain time lag (see Sect. 4.6.4). An issuer's commitment to his business strategy and his Pfandbrief program (see Sect. 4.6.2) might provide some ease in this context.

Maintenance of Overcollateralization

Overcollateralization acts as a buffer against cover pool risks and is an important form of credit enhancement. The issuer's decision regarding the actual level of voluntary overcollateralization maintained is driven by his general business strategy and the target Pfandbrief rating. To obtain this target rating, rating agencies typically require a certain level of overcollateralization to be held, which exceeds legal requirements and depends, among others, on the target rating itself and the cover pool's quality. Overcollateralization is also impacted by funding needs: the Pfandbrief bank might, for example, decide to issue further Pfandbriefe without posting additional cover assets, or withdraw assets from the cover pool to use them for other forms of collateralized funding. Voluntary overcollateralization may be influenced by other drivers, too. Figure 22 summarizes the drivers that impact the maintenance of voluntary overcollateralization.

The main issue in the context of voluntary overcollateralization is its voluntary nature: there are almost no restrictions for the issuer to suddenly reduce it. This becomes especially relevant in a situation of financial distress, when the issuer might have no other choice but to reduce voluntary overcollateralization to ease funding pressure. Another problem is that, in practice, reasons for fluctuations in overcollateralization levels are not always apparent, as for example in the case of redemptions or ongoing issuance of registered Pfandbriefe (S&P 2007c). This makes it hard for the Pfandbrief investor to assess the risks associated with the maintenance of overcollateralization.

According to the PfandBG, the cover pool monitor has to give his consent to a deregistration of cover pool assets. The extent to which cover pool assets can be reduced would, therefore, depend on him and the criteria he applies for this judgement. The incentive of issuers to support Pfandbrief ratings by maintaining voluntary overcollateralization had always been considered to be very strong, but this view has changed in the recent past. A survey conducted by S&P in 2007 showed that covered bond holders typically did not consider the voluntary overcollateralization needed to maintain the current rating to be an unencumbered asset in their liquidity stresses (S&P 2007c). Furthermore, according to the German Government, Allgemeine Hypothekenbank Rheinboden AG did always maintain the overcollateralization levels required by the rating agencies when in distress (Deutscher Bundestag 2006), and even Hypo Real Estate AG, in preparation for government support in September 2008, did not reduce the Pfandbrief cover pools to collateralize new borrowings (Deutscher Bundestag 2009c, p. 238). Still, as recent developments show, the maintenance of high levels of voluntary overcollateralization cannot be relied upon: One (but not the only) example is BBVA, a Spanish covered bond issuer, which in 2011 demonstrated that—even if there were no large fluctuations in the level of overcollateralization in the past, and the issuer had never reduced the level of overcollateralization to a large extent—overcollateralization can still be suddenly reduced to the legally required minimum level (Commerzbank 2011c). In BBVA's case, the reduction of overcollateralization was a consequence of considerable amounts of new covered bond issuances. There may, however, also be other reasons for the reduction of overcollateralization, such as a withdrawal of overcollateralization due to reduced issuer support to the Pfandbrief program, which is discussed in more detail in Sect. 4.6.2. Rating agencies do not always give full credit to voluntary overcollateralization (see Moody's 2012c; Fitch 2012c), and in case of low issuer ratings or short term ratings they tend to recognize voluntary overcollateralization only if it has been assured by contract or public statement (Commerzbank 2011a).[27] Programs which are dormant or in wind-down are seen particularly critical (Fitch 2012c). Results obtained by Siewert and Vonhoff (2011) also confirm that the amount of available overcollateralization does not have a big impact on Pfandbrief spreads, i.e. investors do not seem to take into account this type of overcollateralization when pricing a Pfandbrief.

The question regarding the overcollateralization available at the time of issuer default is of high importance to the Pfandbrief holder. It is, however, hard to predict how fast and by how much the issuer would actually be able (and allowed)

---

[27] Moody's (2012c), for example, distinguishes between committed and uncommitted overcollateralization. While committed overcollateralization (which the issuer is obliged to maintain, either by legislation or contractual terms) is fully accounted for, the value given to uncommitted overcollateralization depends on several factors such as the issuer's rating and the strength of the covered bond's legal or contractual framework. So far, two German issuers (Bayerische Landesbank and SEB) have committed to an overcollateralization above the legally required one (cf. Moody's 2013a).

to reduce the overcollateralization before he goes into default. Therefore, it should not be assumed that a current level of voluntary overcollateralization will still be fully available upon issuer insolvency. In a worst case scenario, the issuer could reduce overcollateralization to the legally required amount shortly before he goes into default.[28]

4.1.3 Further Pfandbrief Issuance

The issuer's decision to issue further Pfandbriefe can have a negative impact on existing Pfandbriefe as it might reduce the available overcollateralization if no additional cover pool assets are posted. In the event the cover pool is increased simultaneously, its overall quality might suffer if the additional assets are of lower quality. The asset liability matching profile may also weaken, for example if the additional liabilities are short-dated large-volume Jumbo Pfandbriefe. As outlined by Fitch (2008), excessive reliance on Pfandbrief funding can result in risk concentrations on the issuer's asset and liability side, and in a decline of unencumbered assets available to unsecured and subordinated creditors that may affect the issuer's senior unsecured rating. This could, in turn, have an impact on the Pfandbrief rating, and would be worsened in case of excessive cherry-picking (i.e. taking good quality assets into the cover pool and thereby reducing the quality of assets available to the obligations of unsecured creditors). As pointed out by Berninger and Winkler (2012), the issuer's reliance on covered bond funding depends on his business model and is likely to be higher for specialized lenders than for universal banks.

The PfandBG restricts the issuance of new Pfandbriefe only by the cover requirements discussed in Sect. 2.8, but apart from that there is no limit regarding the volume of outstanding Pfandbriefe. The latest amendment of the PfandBG requires the disclosure of the amount of cover pool assets which are, in principle, cover pool eligible but exceed certain limits imposed by the PfandBG (Bundesrat 2013, p. 73). This will give valuable information on how much room there is for additional issuances. According to Berninger and Winkler (2012), the ratio of cover pool assets to total assets is comparably high for specialized lenders in Germany.

In 2008, Fitch (2008) considered the downside risk of (further) covered bond issuance to be limited. However, discussions about *asset encumbrance*, i.e. the reservation of certain assets on the bank's balance sheet for specific creditors, have gained importance recently, mainly in the context of the European Comission's proposed EU directive on a framework for recovery and resolution of credit institutions (European Commission 2012), which is discussed in more detail in Sect. 4.5.5. The main question in the context of asset encumbrance is how many assets will, in case of a bank default, still be available in the general insolvency estate, i.e. for the remaining creditors (such as senior unsecured creditors). Asset encumbrance does, however, not only relate to covered bonds, but also to central bank funding, derivative transac-

[28] The cover pool monitor who gives his consent to a deregistration of assets makes sure that the mandatory overcollateralization is not undercut.

tions, other types of secured funding and securitization, and needs to be considered on an overall level. In the context of covered bonds, it is aggravated by the increasing overcollateralization requirements from rating agencies (Volk and Will 2012). In newer covered bond jurisdictions, such as Belgium, Korea, New Zealand, Australia and Singapore, regulators have defined fixed thresholds for covered bond issuance. In the UK, issuance is also restricted, but a case-by-case consideration applies (Moody's 2012a). As pointed out by Moody's (2012a), such an issuance restriction can have positive and negative effects on the issuer's credit quality. For more details on the topic of further Pfandbrief issuance see, for example, Fitch (2008, 2011b,e). Further information on asset encumbrance can be found in Berninger and Winkler (2012); Stöcker (2012) and Volk and Will (2012).

4.1.4 Other Issuer-Related Risks

There are also other issuer-related risks that are relevant to the Pfandbrief holder, such as operational risk, reputation risk and strategic risk.

Operational Risk

The risk of losses caused by the issuer's daily business activities, such as peoples' actions (e.g. mistakes, deficient qualification, fraud/theft), systems (e.g. system outage, bugs, data security), inadequate processes (e.g. lack of control, disruption of business process) or external events (e.g. delicts of third parties, natural hazards, law amendments) is called *operational risk*, see Wolke (2008), p. 202. It also includes origination, underwriting and servicing operations. As it strongly depends on the respective issuer, it is difficult to assess for an external investor. Operational risk is not Pfandbrief-specific and inherent in many financial products. It cannot be completely ruled out.

Reputational Risk

Reputational risk refers to the trustworthiness of the Pfandbrief issuing bank. An adverse perception of the issuer in the market may have a negative impact on the Pfandbrief. Similar to operational risk, issuer reputation risk is inherent in many financial products and also cannot be completely ruled out.

Strategic Risk

The risk of losses arising from strategic business decisions of the issuer is referred to as *strategic risk*. Again, strategic risk is inherent in many financial products and cannot be completely ruled out either.

4.1.5 Conclusion

Issuer risk plays a crucial role as it is the risk the Pfandbrief holder is exposed to in the first place: Issuer default triggers the Pfandbrief holder's direct exposure to cover pool risks, the risk of timely Pfandbrief redemption and structural and legal risks, while cover pool management influences both the cover pool composition and the overcollateralization available at the time of issuer default. Similarly, further Pfandbrief issuance under distress may reduce available overcollateralization. The risk of further Pfandbrief issuance (apart from the impact on overcollateralization levels) and other issuer-related risks as discussed in this section should be of secondary importance only in the context of Pfandbrief-related risks.

Several Pfandbrief bank bailouts in the past have highlighted the relevance of issuer default risk, and the recent financial crisis has shown that the market's perception of default risk increases in times of stress. Furthermore, in light of the eurozone sovereign crisis, the credit quality of some German issuers has deteriorated. The event of issuer default itself does, however, not automatically cause losses to the Pfandbrief holder as there is still recourse to the cover pool (and the issuer's general insolvency estate). Actual losses to the Pfandbrief holder depend, among others, on the cover pool composition and the overcollateralization available upon issuer insolvency, and on the extent to which the risks discussed in Sects. 4.2–4.5 materialize.

4.2 Cover Pool Risks

As long as the issuer is solvent, the Pfandbrief holder is not exposed to cover pool risks, as they are absorbed by the issuer. Upon issuer insolvency, however, the cover pool is the Pfandbrief holder's only source of payments and, therefore, its performance becomes important for him. Deteriorating quality of cover pool assets or defaults of derivative counterparties might cause losses to the cover pool (credit risk). Moreover, mismatches in interest rates and currencies may result in mark-to-market losses, or cause cover pool cash flows to be insufficient to make scheduled payments (interest rate and currency risk). In addition to that, mismatches in the maturity profile—with Pfandbrief maturities being shorter than asset maturities—cause refinancing risk in the cover pool, especially in case of large *bullet maturities*. Reinvestment risk arises from maturing, prepaying or defaulting assets. In the case of Mortgage Pfandbriefe, the Pfandbrief holder is also exposed to real estate risk.

4.2.1 Credit Risk

Credit risk refers to losses arising from the cover pool due to changing credit quality of its assets (asset credit risk) or derivative counterparties (counterparty credit risk). The actual credit risk exposure depends on the cover pool's overall composition, as well as the type and quality of its assets and derivative counterparties.

Fig. 23 Mitigation of asset credit risk through the PfandBG's provisions

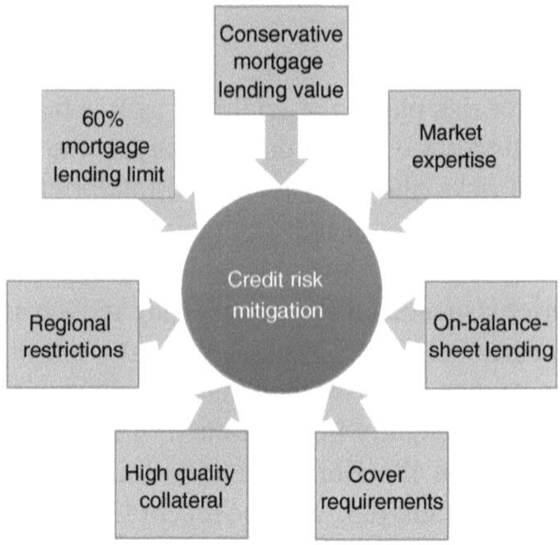

Asset Credit Risk

Deteriorating credit quality of cover pool assets can lead to defaults or to mark-to-market losses when an asset needs to be sold (*asset credit risk*). A rating downgrade—which reflects a higher default probability—may also affect the cover pool to the extent that it has an impact on the asset's cover pool eligibility[29] or on its realisable market value (*market value risk*). Asset credit risk does, therefore, not only comprise asset default risk, but also downgrade risk and, more generally, *credit spread risk*. While not relevant from a cash flow perspective, the effects of changing credit spreads which are not due to worsening credit quality but due to changed risk aversion in the market still play a role in the context of refinancing and reinvestment activities (see Sects. 4.2.4 and 4.2.5). Asset defaults can also disrupt matched cash flows, introduce additional asset liability mismatches, or result in reinvestment risk (see Sect. 4.2.5) if the asset's recovery value is paid prior to the asset's legal maturity. The credit risk characteristics of a cover pool asset (such as default probability and recovery rate) largely depend on the specific type of collateral. According to Moody's collateral score, the credit quality of German public sector cover pool assets is, for example, on average higher than for German mortgage cover pool assets (see Moody's 2013a and previous publications). Furthermore, commercial mortgage assets are typically considered to be more risky than residential mortgage assets (cf. Dübel 2010).

The PfandBG stipulates high quality requirements for cover pool assets, and the range of eligible collateral is quite restrictive. MBS, student loans and other ABS are, for example, not allowed to be included in the cover pool, and there are also regional restrictions. As compared to other jurisdictions, mortgage lending is very conservative, and the 60 % limit on the mortgage lending value provides an important cushion

[29] The eligibility criteria require at least credit quality step 2 for certain public sector cover pool assets, cf. Sect. 2.5.

against losses resulting from defaults (see Rudolf and Saunders 2009; S&P 2011).[30] Before assets related to business activities in new markets can be included in the cover pool, the Pfandbrief bank must have obtained the necessary expertise. The fact that cover pool assets remain on the issuer's balance sheet and are, therefore, subject to the same regulatory requirements as the rest of the bank's assets, should also—to some extent—prevent the issuer from excessive risk taking and give an incentive for careful origination and underwriting standards. The 2 % excess cover provides a certain additional cushion against losses resulting from defaults. Figure 23 summarizes all these provisions which limit credit risk arising from cover pool assets to a considerable extent. Additional voluntary overcollateralization can further mitigate it. Even though asset credit risk is limited to a considerable extent, it is still present:

- The buffer provided by the 60 % mortgage lending limit is itself subject to risk (real estate risk). For more details, see Sect. 4.2.7.
- As pointed out by Packmohr (2011), there is no restriction with respect to the creditworthiness of EU sovereign debt, and countries that are subject to default or restructuring do not have to be excluded from the cover pool. While mortgage cover assets still benefit from a claim against a real estate property in case the debtor defaults, public sector cover assets do not offer an additional source of recourse, and debt repayment depends to a certain extent on the political willingness to do so (Dierks and Engelhard 2010). In the past, sovereign debt was considered to be (nearly) risk-free, and so this was not a big issue. In view of the eurozone sovereign debt crisis and the related increase in sovereign risk (see Sect. 4.5.1), however, this topic has become more relevant, and rating agencies now have a closer look at public sector collateral (Fitch 2013a), also in the case of German Pfandbriefe (Moody's 2012h). Not only Public Sector Pfandbriefe are affected by sovereign risk, but also Mortgage Pfandbriefe, as these allow—to a certain extent—the inclusion of certain public sector bonds. Public Pfandbriefe show high concentrations in the German market as can be seen from the § 28 reporting statistics published on the vdp's website (vdp 2013e): as of Q4 2012, 81 % of public sector assets were from Germany, while the exposure to Spanish, Italian, Portuguese and Irish debtors only accounted for around 5 % (there were no Greek or Cypriot assets left). Furthermore, 77 % of all mortgage cover pool assets were German assets. To further ease the concerns about sovereign risk, vdp member banks have agreed on the so-called *vdp credit quality differentiation model*, which goes beyond legal requirements and accounts for differences in the sovereigns' creditworthiness. Since end of February 2013 (Q4 2012 reporting), vdp member banks apply these standards on a voluntary basis, on top of the normal cover pool reporting (vdp 2013b, p. 4). Maturity-independent haircuts are applied to the nominal value of public sector claims from European countries where the corresponding central state has a non-investment grade rating. The size of these haircuts is based on the 10-year cumulative default probabilities incorporated in the corresponding ratings from Moody's, S&P and Fitch, i.e. lower ratings get

[30] In other jurisdictions, loan-to-value limits can be up to 80 %, and the quality of the property valuation also varies (Volk 2011a).

a higher haircut. For more details on the vdp credit quality differentiation model, see for example Tolckmitt and Stöcker (2012b), or DG Hyp (2012).
- The PfandBG does not limit *concentration risk*, and as just mentioned German Pfandbriefe typically show high concentrations in the German market. Concentration risk might become especially relevant to Public Pfandbriefe should the credit quality of the German government deteriorate. On the other hand, as long as this is not the case, such high concentrations can also turn out to be extremely valuable. The fact that concentration risk is disregarded by the PfandBG is also pointed out by S&P (2004b), and Dübel (2010) explicitly suggests a limitation of concentration risks in the public sector Pfandbrief business. In some other jurisdictions, there are concentration limits to constrain exposures on single issuers to a defined percentage of the cover pool (ECB 2008).
- The buffer resulting from the excess cover requirements is not sensitive to changes in credit spreads (Packmohr 2011). Credit spreads, in turn, determine the realizable sales price in case of forced asset liquidation. As pointed out by Commerzbank (2011b), in the context of sovereign risk and rising sovereign spreads it might be questionable to what extent the cover pool's excess cover resulting from the PfandBarwertV is still adequate. Packmohr (2011) discusses alternatives.
- Further cover assets, which may contain claims against the public sector and suitable credit institutions, currently have to be disclosed with their notional value only, meaning that not much is known about the actual composition of this position (e.g. with respect to debtors, ratings, countries). As opposed to the case of mortgage cover assets, there is no additional source of recourse in the form of a claim against a real estate property for these assets, implying a lower recovery in case of default. As can be seen from § 28 reporting publications as of Q4 2012 (see vdp 2013e), the amount of such further cover assets is not negligible, especially for Mortgage Pfandbriefe. The issue has been addressed by the latest amendment of the PfandBG, which requires more detailed information on further cover pool assets to be disclosed (see Bundesrat 2013, p. 73).

All in all, under normal market conditions, asset credit risk arising from Pfandbrief eligible mortgage and public sector lending is mitigated to a considerable extent, which is supported by the results obtained by Sünderhauf (2006). In times of financial distress, however, asset credit risk can become a big issue. Siewert and Vonhoff (2011) find that the quality of cover assets is less relevant under normal market conditions. During times of economic distress, however, the quality of the cover pool (e.g., in their model, the fraction of German cover pool assets or the extent of term transformation) can have a significant impact on Pfandbrief spreads.

Counterparty Credit Risk

The need to rehedge the cover pool after a derivative counterparty's default may cause losses if the market has moved in an adverse manner and hedging has become more expensive. In case the derivative had a positive market value, the cover pool is hit by a corresponding mark-to-market loss, even if there is no need to rehedge. *Counterparty (credit) risk* is especially relevant when there are high counterparty concentrations, when the derivative counterparties are somehow related to the Pfandbrief issuer or when there is a high default correlation between both.

The PfandBG limits the derivative exposure to 12 % on a net present value basis. When this limit is reached, the Pfandbrief issuer can increase the overcollateralization or reduce the claims/liabilities due under the derivative contracts in the cover pool. There is also the 2 % excess cover that provides some additional buffer. Counterparty credit risk can be further mitigated by collateral agreements with the derivative counterparties, or by additional voluntary overcollateralization. Nowadays, the majority of OTC derivative positions is collateralized (International Swaps And Derivatives Association 2010) and, according to Fitch (2012b) this also applies to most derivatives used in covered bond programs. In practice, only few Pfandbrief issuers actually have derivatives in their cover pools (Schulz 2012). Counterparty credit risk should, therefore, be low as compared to other cover pool risks.[31]

4.2.2 Interest Rate Risk

Interest rate risk refers to losses due to adverse interest rate movements. These losses might be caused by interest rate mismatches between the cover pool and outstanding Pfandbriefe, or by mark-to-market losses in case of a forced sale. A Pfandbrief paying a fixed (floating) coupon with a comparably high proportion of floating (fix) rate assets in the cover pool might experience problems in a low (high) interest rate scenario, when interest rate payments received from the cover pool are not sufficient any more to generate enough proceeds to make interest rate payments due on the liabilities. In a forced-sale-situation, a lower realisable price caused by unfavourable interest rate movements would have a negative impact on the cover pool's value. The extent to which the Pfandbrief holder is exposed to interest rate risk depends on the size and the duration of interest rate mismatches.

According to Fitch (2010), interest rate mismatches are common in Pfandbrief programs and typically addressed by *natural hedging* and overcollateralization, but not so much by privileged interest rate swaps. The PfandBG does not directly require interest rate risk to be hedged, but its provisions regarding cover on a stressed net present value basis reduce interest rate risk considerably (Fitch 2004). They do, however, only account for moderate interest rate stresses i.e. they do not ensure that the cover pool would be able to withstand extreme market conditions (S&P 2004b). The latest amendment of the PfandBG requests the publication of the share of fix rate cover pool assets and fix rate Pfandbriefe (Bundesrat 2013, p. 73), which will provide valuable additional information in the context of interest rate risk.

---

[31] As of May 2009, only 2 out of 13 German Mortgage Pfandbrief programs considered by Fitch (2009a) used privileged derivatives. For Public Pfandbriefe, it was 6 out of 17. § 28 reporting data on the vdp's website (vdp 2013e) shows that the situation has not changed materially: as of Q4 2012, the majority of issuers still did not use derivatives.

4.2.3 Currency Risk

The risk of losses due to adverse movements in exchange rates is referred to as *currency risk*. Currency mismatches between the cover pool and outstanding Pfandbriefe or mark-to-market losses in case of a forced sale can cause such losses. When a high proportion of cover pool assets (outstanding Pfandbriefe) is denominated in another currency and that currency is depreciated (appreciated), cover pool proceeds may not any more generate enough proceeds to make payments to covered bond holders. The extent to which the Pfandbrief holder is exposed to currency risk depends on the size and duration of the mismatches.

As opposed to other legislations, the PfandBG does not contain any provisions that require currency mismatches to be hedged. Therefore, similarly to the case of interest rates mismatches, currency mismatches are common in German Pfandbrief programs, privileged currency swaps are not frequent, and currency risk is mainly addressed via natural hedging and overcollateralization (Fitch 2010; Schulz 2011). Again, the PfandBG's provisions regarding cover on a stressed net present value act as risk mitigant (Fitch 2004). Furthermore, outstanding Pfandbriefe are to a large extent denominated in euros, while cover pool assets show high concentrations to borrowers in Germany and euro countries (see § 28 reporting statistics on the vdp's website, vdp 2013e).[32] Therefore, currency risk is limited to a considerable extent, but as in the case of interest rate risk it should be noted that the stressed net present value only takes into account moderate stress scenarios and does not ensure that the cover pool would be able to withstand extreme market conditions. The latest amendment of the PfandBG will provide additional information in the context of currency risks as it includes disclosure requirements with respect to the net present value of assets and liabilities per foreign currency (Bundesrat 2013, p. 73).

4.2.4 Refinancing Risk

Refinancing risk arises from maturity mismatches between assets and liabilities when the natural amortization of the cover pool is not sufficient to match the maturity profile of the outstanding Pfandbriefe. Typically, the maturities of cover pool assets are longer than the maturities of the corresponding covered bonds (Soldera et al. 2012). Mismatches can also be caused by atomistic and amortizing assets as opposed to covered bonds in bullet format and/or large sizes (Hillenbrand 2011). As a consequence, additional funds might have to be raised by the cover pool administrator to ensure timely payments to privileged creditors after issuer insolvency. The extent to which the cover pool is exposed to refinancing risk depends on the maturity mismatches, the cover pool administrator's range of funding possibilities, the time restrictions he has to cope with, and the conditions to which funds can be raised. The conditions to which funds can be raised, in turn, depend on the stability of the local financial mar-

[32] As outlined in Sect. 2.5, there is a 10 % limit for debt from outside the EU, and the euro is the official currency in 17 out of the 27 EU member states (European Commission 2013).

kets in general, the particular market environment at that point in time, and the type and quality of the cover pool assets. While traded bonds might be sold comparably easily, it could be more problematic to sell mortgage loans.[33] Near-term maturity mismatches are typically seen more critical than those occurring in the medium or long term as the cover pool manager will have less time for action. Refinancing risk is especially high in the case of large-volume Jumbo Pfandbriefe becoming due shortly after issuer default. In a stressed market environment (such as in the immediate aftermath of an issuer default), it might be difficult to raise funds and asset sales may only be possible at large discounts (market value risk), especially when the market is aware of the cover pool's refinancing pressure. In this context, sovereign risk (see Sect. 4.5.1) and the strength of the financial system in general play an important role (Moody's 2012c). The (local) sovereign is often considered as lender of last resort, and Moody's (2011b) regards the price of government debt as a lower bound for refinancing costs.

The 180-days liquidity buffer mitigates short term refinancing risk to a considerable extent as it requires liquidity gaps within the next 6 month to be covered by highly liquid assets. This gives the cover pool administrator sufficient time for action and reduces the likelihood of fire sales being necessary shortly after issuer default. In addition to that, the PfandBG allows for the inclusion of further cover assets to increase cover pool liquidity. The cover pool administrator also has a broad range of possibilities to raise liquidity against the cover pool (see Sect. 2.12). As a consequence, S&P assigns the best category regarding range and strength of funding options to German Pfandbriefe (S&P 2009, 2012a). The mandatory overcollateralization also provides a certain cushion against losses from refinancing risk, and it can be further mitigated by additional voluntary overcollateralization (especially in the form of liquid collateral) or by natural hedging.

While the provisions of the PfandBG mitigate refinancing risk to a considerable extent, they do not eliminate it. The liquidity buffer calculations only look at projected cash flows, but they do not address liquidity gaps arising from cash flow interruptions (i.e. from non-performing assets). They also ignore the fact that assets used to cover liquidity gaps might themselves be exposed to some level of market value risk, although the volatility should be low due to the nature of these assets. Most importantly, mismatches arising after the 180-days period are ignored, and the 2 % excess cover does not necessarily ensure that cash inflows are sufficient to cover cash outflows at all future points in time. In this context, the latest amendment of the PfandBG, which includes disclosure requirements on the maturity profile of outstanding Pfandbriefe and fixed interest rate periods in 6 month bands for the first 2 years (Bundesrat 2013, p. 73), will provide valuable additional information on the cover pool's refinancing situation.

[33] This is, for example, reflected in the average refinancing margins used by Moody's for German Pfandbriefe (Moody's 2012c). These refinancing margins reflect the discount that a purchaser would require when buying the cover pool assets. They are normally lower for public sector assets than for mortgage assets. Fitch also considers mortgages to be less liquid than public sector assets (Fitch 2012d).

Refinancing risk is considered to be "extremely volatile" (Forster et al 2011), and due to the financial crisis, it has gained importance over the last years. It is also in the focus of rating agencies, who have adjusted their methodologies, among others with respect to assumptions on market liquidity and liquidation prices post-issuer insolvency. Moody's (2011a) considers the refinancing risk of most covered bonds to be significant. However, for German Pfandbriefe Moody's quarterly published losses from market risk, the main component of which is refinancing risk (cf. Rudolf et al. 2012), have—on average—remained stable from Q3 2009 to Q3 2012, and as of Q3 2012 Moody's "Timely Payment Indicator" assigns a high probability of timely payment upon issuer insolvency to all German (Mortgage and Public) Pfandbriefe issued under the PfandBG, see *Moody's EMEA/European Covered Bonds Monitoring Overview* (Moody's 2010e,f,g,h, 2011c,d,f, 2012d,g,e,f, 2013a).

In other covered bond markets, refinancing risks are mitigated by *pass-through structures* and by covered bonds with *soft bullet maturities* (see Volk 2011a). While covered bonds with pass-through structures are repaid from the natural amortization of the assets, structures with soft bullet maturities allow, under certain conditions, the repayment of covered bonds to be delayed while coupon payments are continued and may help to reduce refinancing pressure.[34] For more details on the assessment of asset liquidity and refinancing risks for covered bonds from the perspective of a rating agency see, for example, Fitch (2012d, 2013d). Additional information on the current discussions on timely payments and soft bullet/pass-through structures can be found in Rudolf et al. (2012).

4.2.5 Reinvestment Risk

Reinvestment risk refers to the uncertainty regarding the conditions to which cover pool proceeds can be reinvested when they are not needed to make Pfandbrief payments. It arises from maturity mismatches when assets are being repaid earlier than Pfandbriefe, from asset prepayments, or from defaults with recoveries. Reinvestment risk becomes relevant to the Pfandbrief holder after issuer default when the lower yield of newly added assets results in a *negative carry*. If this happens to a large extent, interest proceeds from the cover pool might not be sufficient any more to make interest payments to Pfandbrief holders. Reinvestment risk is especially high when the majority of cover pool assets has shorter maturities than the outstanding Pfandbriefe.

The PfandBG's excess cover requirement provides a certain cushion against reinvestment risk, and it can be further mitigated by additional voluntary overcollateralization, or by natural hedging. As will be discussed in Sect. 4.2.6, prepayment risk is comparably low, i.e. reinvestment risk mainly arises from maturing or defaulting assets. The weighted average life of cover pool assets is typically longer than the weighted average life of the corresponding covered bonds (Soldera et al. 2012), which

[34] The exact wording of the extension option, however, plays an important role. For more details, see Hillenbrand (2011).

further mitigates reinvestment risk from maturing assets. In case of asset defaults, the question regarding the notional repayment should be more material than the lower interest earned on the reinvestment of the assets' recovery value.

4.2.6 Prepayment Risk

Prepayment risk arises from the early repayment of cover pool debt. Normally, prepayments in the cover pool are absorbed by the Pfandbrief issuer. After issuer insolvency, however, the Pfandbrief holder is exposed to prepayments risk, which results in reinvestment risk (see Sect. 4.2.5). In Germany, actual mortgage loan prepayment rates are low (Volk 2011a, p. 77). This is mainly due to legal restrictions which can be summarized as follows (cf. Rudolf and Saunders 2009, p. 18).

> [L]oans typically exhibit a fixed interest rate and usually do not offer a prepayment option to the borrower. An early redemption, if possible at all, will come along with a prepayment indemnity to compensate the lender for potential reinvestment losses. [...] In the first 10 years of a fixed-rate mortgage, the lender normally is not obliged to agree upon an early redemption. Only if the borrower has a legitimate interest in an early redemption, especially when the property is about to be sold, the borrower is granted the right of an early repayment. However, the borrower is still required to compensate the lender for the prepayment. If the fixed-rate period is longer, after 10 years the lender has a legally-defined prepayment option with a six-month notice and without indemnity.

In German public sector lending, prepayment rights can be precluded for more than 10 years (Volk 2011a, p. 77). As German Pfandbriefe show high concentrations in the German market, the relevance of prepayment risk should, therefore, be low in the context of cover pool risks.

4.2.7 Real Estate Risk

Upon issuer insolvency, the Mortgage Pfandbrief holder is also exposed to *real estate risk*, i.e. the risk of losses arising from changing real estate prices. This is due to the fact that the recovery value of a defaulting mortgage asset depends on the proceeds of selling the charged property. In this context, foreclosure costs, time to foreclosure (including accrued interest) and prior ranking mortgages also play a role (Moody's 2009d, p. 12).

Real estate risk only becomes relevant to the Pfandbrief investor in case both the Pfandbrief bank and the mortgage debtor default. Conservative mortgage lending as stipulated by the PfandBG mitigates real estate risk:

- The determination of the mortgage lending value is itself very conservative, resulting in mortgage lending values being on average 10 %, but sometimes up to 20 % lower than the market value (Fitch 2011d). In case of booming foreign sub-markets, differences of more than 40 % were in some cases observed in the past (vdp).
- The 60 % mortgage lending limit provides an additional buffer that reduces the impact of market value changes in the underlying properties on the Pfandbrief

holder's claim: Losses on a mortgage loan would only occur if the borrower defaulted and the proceeds from the foreclosure were less than 60 % of the property's value (S&P 2011).

- A regular review of the mortgage lending value and specified revaluation triggers make sure that considerable changes in real estate prices are detected and lead to an adjustment (and a reduction of the corresponding loan's cover pool eligible amount) if necessary. This might, for example, happen in case of "far-reaching and, in all likelihood, prolonged market disruptions", as observed in certain markets during the financial crisis (Dresch 2010).

Even in case of a considerable decline in real estate markets, the Pfandbrief holder is, therefore, not necessarily affected by losses in case of the mortgage debtor's default. The fact that German property markets (which typically constitute a large share of Pfandbrief cover pools) are considered to be very stable, and that their volatility is comparably low by international standards (Berninger 2010) provides additional ease in this context. However, the housing market in Germany is currently in a growth phase. In some markets, prices and rents have already increased considerably. According to the vdp, no warning signals can be observed at the moment (cf. vdp 2013f): the developments are plausible and they can be fundamentally explained. Yet, the whole development should be kept in view.

All in all, the PfandBG's provisions with respect to mortgage lending result in a high level of comfort that, under normal market conditions, real estate risk does not affect the Pfandbrief holder to a large extent. The PfandBG can, however, not rule out the impact of a systematic overvaluation of properties or a massive decline in property prices with a simultaneous increase in the default rate of borrowers.

4.2.8 Summary

The above analysis of cover pool risks suggests that refinancing risk, market risk (mainly interest rate, but also currency risk) and asset credit risk are the major cover pool risks the Pfandbrief holder is exposed to. Reinvestment risk, prepayment risk and counterparty credit risk are less important. Real estate risk only becomes relevant upon the mortgage debtor's default and is, therefore, not considered explicitly in the following, but as a part of asset credit risk (as it finally determines the loss severity of the corresponding mortgage asset).

These findings are supported by S&P (2011), who consider asset liability mismatches and cover pool credit risk (in that order) to be the most important sources of risk for Pfandbrief holders. Moody's Pfandbrief cover pool losses from *market risk* (which here includes refinancing risk, interest rate and currency risks) are typically higher than cover pool losses from *collateral risk* (i.e. credit risk), see *Moody's EMEA/European Covered Bonds Monitoring Overview* (Moody's 2010e,f,g,h, 2011c,d,f, 2012d,g,e,f, 2013a). This suggests that, for German Pfandbriefe, refinancing risk and market risk (from interest rate and currency mismatches) are seen more critical than asset credit risk, and is supported by S&P statistics indi-

Fig. 24 Decreasing relevance of cover pool risks post-issuer insolvency

cating that overcollateralization requirements are mainly driven by market risk, and not by credit risk (Volk 2011c). Refinancing risk gets the highest relevance. This is due to the fact that maturity mismatches are inherent in every Pfandbrief program. In case of forced asset liquidation caused by refinancing pressure credit spread risk, as well as interest rate and currency risk, would also materialize as a consequence from market value risk. Figure 24 summarizes all these findings.

4.2.9 Conclusion

As cover pool risks arise from the nature of the Pfandbrief business, there will always be a trade-off between the risks associated with the cover pool on the one hand, and the quality of the Pfandbriefe on the other hand: To better protect Pfandbrief cash flows, the cover pool needs to generate extra revenues which is often achieved by including riskier assets or generating revenues from interest rate or maturity mismatches.

The detailed regulations of the PfandBG regarding cover pool eligibility and cover requirements mitigate individual (i.e. program-specific) cover pool risks to a considerable extent. However, like other covered bond laws, the PfandBG cannot rule out the impact of systemic risks such as a collapse of the (local) financial system or a general deterioration of cover pool asset quality. Under extreme market conditions, cover pool risk (notably refinancing risk, interest rate risk, currency risk and asset credit risk) can, therefore, become a big issue. In the context of cover pool risks, overcollateralization plays an important role as credit enhancement. The high quality of Mortgage Pfandbriefe is partially owed by the stability of the German real estate market, while German public sector cover pools clearly benefit from the creditworthiness of the German sovereign. Should these prerequisites change, concentration risk might become an issue.

The lower the creditworthiness of the Pfandbrief issuer, the more important the quality of the cover pool becomes. Risks inherent in the cover pool should, however, already be considered carefully before issuer default, as they might also have an impact on the issuer's solvency. All in all, the long-term quality of a specific Pfandbrief cover pool is difficult to assess as it requires detailed and up-to-date information on the cover pool and the bank's future strategy.

4.3 Risk of Timely Pfandbrief Repayment

The *risk of timely Pfandbrief repayment* refers to losses arising from the Pfandbrief being repaid earlier or later than the contractual maturity date.

4.3.1 Early Redemption Risk

The risk that the Pfandbrief holder is repaid earlier than the contractual maturity date is referred to as *early redemption risk.* In case of early redemption, the Pfandbrief holder may not be able to reinvest the nominal at the same conditions for the remaining time to contractual maturity and earn a lower coupon only.

In contrast to senior unsecured debt, Pfandbriefe do not automatically accelerate in case of issuer insolvency. As long as the issuer does not have a call right, early repayment can only occur when a cover pool insolvency proceeding has been initiated (Volk 2011a, p. 78). This is, however, already the worst case scenario (i.e. both issuer and cover pool insolvency) in which the risk that the cover pool is not sufficient to repay the full Pfandbrief nominal is likely to be more material than foregone interest proceeds.

4.3.2 Extension Risk

The risk that the Pfandbrief holder is repaid later than the contractual maturity date is referred to as *extension risk*. If the Pfandbrief holder urgently needs the Pfandbrief nominal and it is not available to him because the Pfandbrief is paid back late, he might be forced to borrow the money somewhere else under unfavourable conditions. At the contractual maturity date the Pfandbrief holder might also have the possibility to invest the money at more favourable market conditions.

As Pfandbriefe typically have *hard bullet maturities*, Pfandbrief extension can only happen as a result of structural risks, or in case of cover pool insolvency when cover pool cash flows are delayed due to a moratorium according to the KWG (Volk 2011a, p. 78). As in the case of early redemption, this is already the worst case scenario in which the risk that the cover pool is not sufficient to repay the Pfandbrief nominal is likely to be more material.

4.3.3 Conclusion

Early redemption and extension can only occur in the worst case scenario (i.e. both issuer and cover pool insolvency), in which the risk that the cover pool is not sufficient to repay the full Pfandbrief nominal is likely to be more material than losses on paid or earned interest rates.

4.4 Structural and Legal Risks

Structural and legal risks refer to the uncertainty regarding the decision of cover pool liquidation, and regarding the effectiveness of the mechanics in place to ensure cover pool separation and transition to alternative management. The availability of a legal (or, as in other jurisdictions, a contractual) framework does not necessarily mean that bankruptcy remoteness and cover pool continuation is also feasible. The whole discussion is, however, rather theoretical as a cover pool's bankruptcy remoteness has not yet been tested in practice.

4.4.1 Segregation Risk

Segregation risk refers to problems occurring in the context of cover pool separation after issuer insolvency. Although asset segregation under the PfandBG is considered to be strong (see, for example, Fitch 2009c), there are still potential issues that might arise in this context, such as uncertainty regarding the enforceability of voluntary overcollateralization and the continuation of cover pool derivatives, or risks arising from cross-border lending, commingling, claw-back and set-off. Figure 25 summarizes these issues.

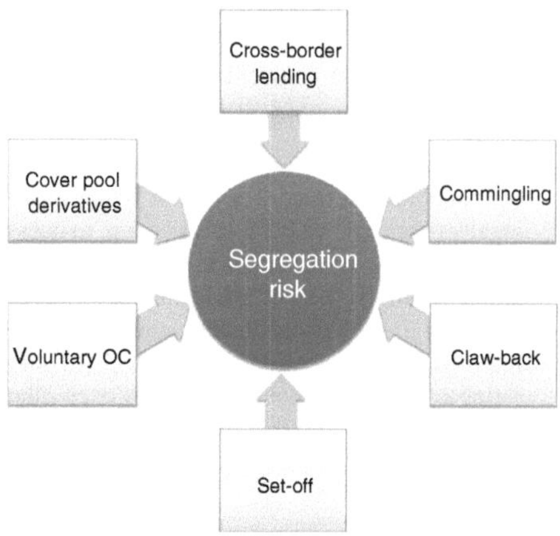

Fig. 25 Potential issues arising in the context of cover pool separation

Enforceability of Voluntary Overcollateralization

The PfandBG does not contain an explicit provision ensuring the bankruptcy-remoteness of voluntary overcollateralization. According to § 30 par. 4 sent. 1 and 2 PfandBG, assets that are "obviously not (...) necessary" as legally required cover, as well as "[a]ssets remaining after the Pfandbrief creditors are satisfied and the management costs are paid", have to be surrendered to the administrator of the bank's general insolvency estate. This conflict of interest between covered bond holders and unsecured creditors is also pointed out by Anhamm (2010): While covered bond holders expect their money at legal maturity (and not before), unsecured creditors want to be paid back as soon as possible upon issuer default. This means that, from the Pfandbrief holder's point of view, it is preferable if overcollateralization is protected until the last covered bond has been paid back. Unsecured creditors, on the other hand, would like to have access to voluntary overcollateralization as soon as possible. If voluntary overcollateralization (or parts of it) were released to unsecured creditors before all Pfandbrief holders had been repaid, Pfandbrief holders could lose credit enhancement otherwise available to them.

There are still remaining doubts regarding the enforceability of voluntary overcollateralization and the interpretation of "will obviously not be necessary". According to Deutscher Bundestag (2009b), "obviously" only applies for disproportionately high overcollateralization, and "will [...] not be necessary" refers not only to the net present value as of one specific cut-off date, but also includes future risks to be considered adequately. A corresponding decision would, therefore, have to be made by a court, and the proof would be up to the administrator of the issuer's general insolvency estate (Deutscher Bundestag 2009b). Moody's (2010d) considers making such a proof as being very difficult in most cases. It is, however, hard to anticipate how a court would decide case-by-case. In the end, the extent to which voluntary overcollateralization is really necessary to fully satisfy the Pfandbrief holder's claims will only unfold when winding up the cover pool. Fitch (2009c) also sees some remaining uncertainty in this context. The German Bundestag's comment on the subject is as follows (Tolckmitt and Stöcker 2012b)[35]:

> A frequent consequence of the stress tests [the rating agencies] apply for the timely servicing of Pfandbriefe is that a Pfandbrief Bank must maintain overcollateralization exceeding the legally required 2 %. In this connection, to question the insolvency remoteness of this 'excess cover' is to ignore the fact that, within the scope of these stress tests, such excess cover would automatically be part of the legally required cover. If a rating agency does not believe that the cover assets are absolutely sound and therefore calculates what overcollateralization is required, then logically these assets that constitute overcollateralization are not (from the rating agency's viewpoint) 'voluntary overcollateralization' at all. Instead, they are part of the (normal) cover, since they are needed (in the rating agency's view) to ensure the Pfandbriefe are serviced on time. They cannot therefore fall within the scope of application of § 30 par. 4 Pfandbrief Act.

---

[35] The original German source is Deutscher Bundestag (2009b).

This does, however, not give a fully satisfying answer, as the overcollateralization required by a rating agency is typically rating-dependent. Therefore, although in practice the whole overcollateralization is often considered bankruptcy remote (see Volk 2011a, p. 78, and Kullmann 2011), there is still some remaining uncertainty with respect to the level of overcollateralization granted to the Pfandbrief holder in case of issuer insolvency.

Continuation of Cover Pool Derivatives

Hedging derivatives that protect the cover pool against interest rate and currency risks are especially important after issuer default. Therefore, it is essential that they continue even if the issuer is insolvent. According to Moody's (2009b) this might, for example, be endangered when issuer default entitles the derivative counterparty to terminate the swap (which is the case under standard, unamended ISDA documentation) or when issuer default results in a failure of the issuer to make payments to the derivative counterparty (which in turn entitles the counterparty to terminate the swap).

As discussed in Sect. 2.5, the PfandBG sets out specific requirements regarding the inclusion of derivatives in the cover pool: They must not contain any termination clause that applies in the case of issuer insolvency. In practice, individual agreements are kept to ensure that netting does not take place between cover pool derivatives and derivatives outside the cover pool. The individual agreement for a cover pool remains in place after issuer insolvency and continues until the respective cover pool becomes insolvent. According to a comment of the Deutscher Bundestag (2009b) it can be assumed that cover pool derivatives survive issuer default as long as the standardized master agreements are used.

Cross-Border Lending Risk

Cross-border lending risk is a legal risk that arises when cover pool assets are located in another country—different from the one the issuing bank is located in—and, therefore, subject to local bankruptcy laws that may not recognize the Pfandbrief holder's priority claim (Volk 2009, p. 14). In this context, comparability of legal positions might not always be straightforward. An unsecured creditor of the issuer could, for example, try to access a foreign asset under secondary proceedings under the respective country's law. The competing claims would then lead to lengthy legal disputes and result in delayed payments to Pfandbrief holders (if the claim can be enforced at all). Mortgage lending is, in general, considered to be more affected by cross-border lending risk than public sector lending (Volk 2009, p. 14).

The PfandBG sets geographical restrictions to account for the fact that extensive foreign lending might result in increased legal uncertainty: for foreign lending outside the EU where the privileged creditors' priority claim cannot be ensured, a 10 % limit applies. For EU countries, there are no restrictions. According to Volk (2011a), p. 22, cross-border lending risk within the EU has been mitigated through the EU directive on the reorganization and winding-up of credit institutions (European Parliament and Council Directive 2001/24/CE), and the vdp has published extensive literature on foreign legal systems and checklists concerning cover pool eligibility of foreign assets (Labe 2010). Risks arising from cross-border lending should, therefore, be comparably low.

Commingling Risk

In case of issuer insolvency, cash flows from the cover assets need to be separated from cash flows belonging to the issuer's general insolvency estate before payments are made. As explained by Fitch (2012b), *commingling risk* refers to the fact that delayed cash flows might cause a temporary interruption of payments to privileged creditors in the immediate aftermath of an issuer default (*commingling liquidity risk*), or lead to a reduction of available overcollateralization if commingled amounts cannot be fully recovered over time (*commingling credit risk*). A delay of cash flows can, for example, occur in the case of loans that only partially belong to the cover pool, or when there are complex and hard-to-handle IT systems. Cash collections made by the issuer and related to cover pool assets might end up in the general insolvency estate and not be available to the cover pool (S&P 2012e). Commingling risk is considered to be higher for commercial real estate and public sector assets than for residential mortgages, which is due to the fact that cash flows arising from residential mortgage assets tend to be scattered over time, while cash flows from public sector and commercial real estate assets are typically more concentrated (S&P 2013).

The PfandBG mitigates commingling risk by a clear definition of assets belonging to the cover pool and by the fact that BaFin can appoint the cover pool administrator already prior to issuer default, which gives him additional time to prepare for the separation of cash flows. Moody's (2010d) expects commingling risk to be limited for German Pfandbriefe.

Claw-Back Risk

Under certain circumstances, the issuer's general insolvency administrator could try to claw back cover pool assets (i.e. to reverse their assignment to the cover pool), which would lead to less cover pool assets being available for the benefit of the Pfandbrief holders.

The PfandBG protects cover pool assets from being clawed-back by registration in the cover register and explicit legal separation. According to Moody's (2010d), for registered cover pool assets there should not be any *claw-back risk* apart from the question regarding the enforceability of voluntary overcollateralization (see above).

Set-Off Risk

Set-off risk is the risk that creditors of the issuer's general insolvency estate attempt to seize cover pool assets to set off other claims they have against the Pfandbrief issuer. Consider, for example, a mortgage borrower who also has a current account with the Pfandbrief bank and wants to set off his claims arising from this current account against his outstanding loan payments. Alternatively, the counterpart of a hedging derivative in the cover pool might try to net payments due under the cover pool derivative against payments owed by the insolvent Pfandbrief issuer under derivative contracts outside the cover pool.

According to Anhamm (2010), set-off is "legally excluded" in Germany, but there is still some remaining uncertainty. Moody's (2010d) excludes set-off risk only for assets that are located in Germany and ruled by German law. For assets with claims against borrowers located outside Germany, or for loans that are not governed by German law, Moody's considers set-off to be a potential risk. In practice, the use of an explicit waiver of set-off in the loan agreements helps to ensure that the cover pool is not reduced as a consequence of set-off activities by the borrower (Stürner and Kern 2007, p. 91).

For German Pfandbriefe, asset-related set-off risk is limited due to the fact that they show high concentrations in the German market. As different individual agreements are used for derivatives, set-off risks in the context of cover pool derivatives are mitigated to a large extent, too, as long as master agreements with different netting sets are used.

4.4.2 Transition to Alternative Management

To ensure timely payments to the Pfandbrief holders in the immediate aftermath of an issuer's insolvency, it is essential that the transition to alternative management happens smoothly, especially when there is a high duration gap and liquidity needs to be raised quickly to ensure cover pool solvency. A smooth transition could, however, be endangered due to a late appointment of the alternative management, missing powers of the alternative manager to take necessary actions, high complexity of the issuers' IT systems (preventing a quick isolation of the privileged assets), or a very complex cover pool that is not easy to administer (Fitch 2009c).

As outlined in Sect. 2.9, the PfandBG regulates the transition to alternative management by clear definitions regarding the appointment of the cover pool administrator and his roles and responsibilities. The cover pool administrator is appointed upon issuer insolvency or even earlier if deemed necessary. This ensures that there will be enough time to enable a smooth transition. Furthermore, all responsibilities for the process post-issuer insolvency are centred around the cover pool administrator, who takes overall responsibility. According to Fitch (2009b), this is preferable to a complex set-up with responsibility being split across different parties. The cover pool administrator is independent from the insolvency administrator and acts on behalf of the privileged creditors only, i.e. there is no risk of conflicting interests. He also

Fig. 26 Issues arising in the context of cover pool liquidation

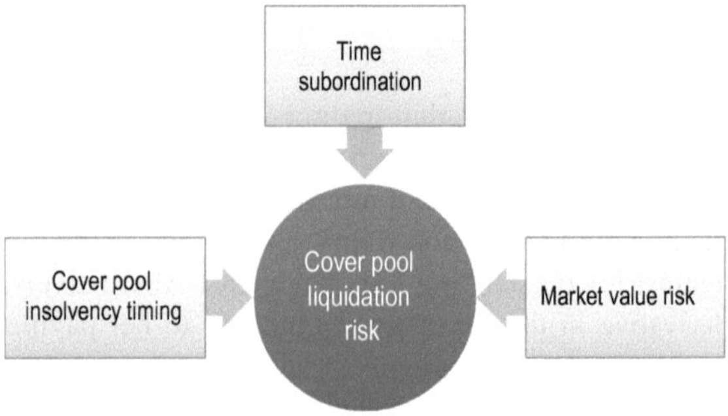

has a clear range of competences to raise funds and acts for a Pfandbrief bank with limited business activity. All in all, as pointed out by Fitch (2009c), the provisions of the PfandBG provide comfort that cover pool management can continue without interruption in case of issuer default.

4.4.3 Cover Pool Liquidation Risk

The PfandBG does not define specific criteria or trigger events regarding the declaration of cover pool insolvency, which means that there is some freedom regarding the decision when cover pool insolvency is to be declared (*cover pool insolvency timing*).[36] Furthermore, there are no restrictions regarding the amount of assets that can be liquidated at once for the repayment of a specific Pfandbrief maturity. This results in repayment risk from *time subordination* for later maturing Pfandbriefe. The gap between an asset's market value and its intrinsic value is an important aspect to be considered in this context (see market value risk, Sect. 4.2). Figure 26 summarizes the above discussed risks which are referred to as *cover pool liquidation risks*.

Consider, for example, a Pfandbrief program with half of the outstanding Pfandbriefe being due shortly after issuer insolvency, and the other half being due 5 years later. In a first scenario, the early maturing Pfandbriefe are paid down using realised proceeds from the sale of available cover assets at large discounts, which results in overcollateralization being reduced significantly, or even completely. In 5 year's time (or even earlier), the cover pool might then turn out to be not sufficient any more to repay the other remaining half of the outstanding Pfandbriefe, i.e. the earlier maturing Pfandbriefe benefit from the decision of cover pool continuation and are not hit by any losses, while the later maturing Pfandbriefe experience losses. In a second scenario, the cover pool is accelerated immediately after issuer default as it is already assumed that the cover pool will not be sufficient to repay all outstanding Pfandbriefe when they are due. All Pfandbrief holders rank pari passu then. As com-

[36] The PfandBG only states that the cover pool administrator must check on a regular basis whether the cover requirements as outlined in Sect. 2.8 are fulfilled, and that there may be additional checks on BaFin's request (§ 31 par. 5 PfandBG).

pared to the first scenario, this might lead to higher losses for the earlier maturing Pfandbriefe, and to lower losses for the later maturing Pfandbriefe.

The minimization of the gap between market value and intrinsic value is an important aspect of cover pool asset management under stress scenarios, but—as pointed out by Engelhard (2010)—the decision whether to sell or retain a specific asset is not always straightforward. While asset liquidation brings in cash and reduces cover pool risks and operational complexities, it might lead to significant discounts to the intrinsic value (market value risk) and eliminates any upside potential due to favourable (future) market conditions. Asset retention, on the other hand, does neither eliminate upside nor downside risk, i.e. the final result may be superior or inferior to (immediate) liquidation, depending on future market conditions.

Moody's (2010d) considers time subordination to be a shortcoming of the German Pfandbrief legislation. Risks from time subordination could be mitigated by *amortization tests* or clauses such as the *SARA clause*, see Fitch (2009b). While amortization tests define clear trigger events for cover pool insolvency which result in cover pool acceleration in case of trigger breaches, the SARA clause prevents the excessive use of overcollateralization for the repayment of earlier maturing covered bonds to the disadvantage of later maturing bonds by allocating overcollateralization to the outstanding covered bonds on a pro-rata basis.

4.4.4 Conclusion

All in all, the PfandBG's regulations to ensure cover pool separation and continuation provide a high level of comfort to the Pfandbrief holder. There is, however, still some uncertainty which is caused by the fact that the PfandBG does not contain an explicit provision that ensures the bankruptcy-remoteness of voluntary overcollateralization. As, until today, no Pfandbrief issuer has defaulted, there is no experience available with respect to the handling of this question. Furthermore, the decision if and when the cover pool (or parts of it) is to be liquidated may be very important for the recovery value of the different outstanding Pfandbriefe. In this context, the Pfandbrief holder is exposed to a considerable amount of uncertainty given that trigger events for cover pool insolvency are not clearly defined, and that provisions limiting the excessive use of overcollateralization for the repayment of earlier maturing Pfandbriefe are missing.

4.5 Other Risks

There are also other Pfandbrief-related risks the Pfandbrief holder is exposed to, including (but not limited to) country risk, operational risk, settlement and related risks, reputational risk and regulatory risk.

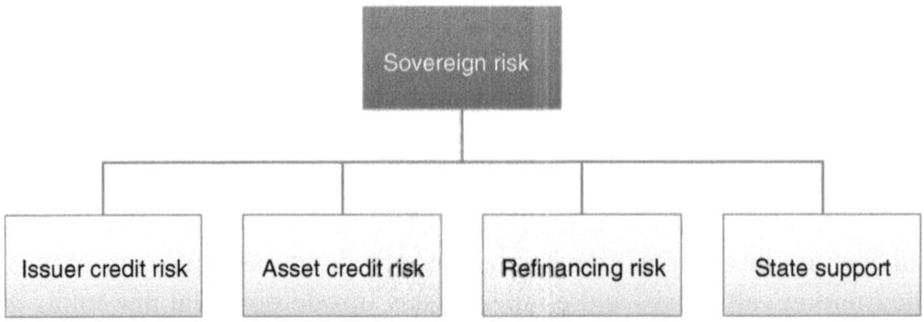

Fig. 27 Role of sovereign risk

4.5.1 Country Risk

The risk of losses arising from a sovereign's inability or unreadiness to provide the required currency to make timely payments is referred to as *country risk*. It is influenced by both international and national factors and includes, among others, sovereign (credit) risk, transfer risk and political risks.

Sovereign (credit) Risk

The risk of losses arising from a government's failure to fulfil its payment obligations in a full and timely manner is referred to as *sovereign (credit) risk*. As pointed out by Packmohr (2011), sovereign credit pressure does not only affect issuer strength (see Sect. 4.1) and cover pool credit quality (see Sect. 4.2.1), but also has an impact on refinancing risks (see Sect. 4.2.4). Furthermore, the strength of the sovereign also influences his ability and willingness of direct or indirect support (see Sect. 4.6.5). This is why bank ratings often account for a certain amount of sovereign support. Figure 27 summarizes the role of sovereign risk in the context of German Pfandbriefe.

Investors increasingly question the ability and willingness of several sovereigns to repay their debt completely and in due time (Dierks and Engelhard 2010) and, since 2010, sovereign risk is driving covered bond spreads (Sarafana 2010; Will 2011; Volk and Will 2012). As sovereign risk impacts the Pfandbrief holder through issuer credit strength, cover pool credit quality, refinancing options and the market's perception of the likelihood of active state support, see the corresponding sections for a respective risk assessment. More details regarding the influence of sovereign credit strength on financial institutions, covered bonds and their ratings can, for example, be found in Moody's (2012b).

Transfer Risk

Losses from *transfer risk* arise when a government does not allow monetary transactions to take place with counterparties in foreign countries, which might prevent mortgage or public sector debtors in such countries to make their scheduled payments. In the context of German Pfandbriefe, transfer risk should be comparably low: The PfandBG sets a limit of 10 % for debt from outside the EU, and cover pools typically show high concentrations in the German market.

Political Risk

The risk of losses due to political changes or instability is referred to as *political risk*. It includes events such as war, revolutions and embargoes. Regarding the relevance of this kind of risk for German Pfandbriefe, the same argumentation as in the case of transfer risk holds.

4.5.2 Operational Risk

Similarly to issuer-related operational risk as discussed in Sect. 4.1.4, there is also *operational risk* arising from all (other) Pfandbrief-related business activities. This includes the activities of the cover pool monitor and, in case of issuer insolvency, the cover pool administrator. Incorrect or forgotten registration of cover pool assets might, for example, result in cover pool assets not being available upon issuer default. Operational risks are inherent in most financial products and cannot be ruled out. As compared to other Pfandbrief-related risks they should, however, be of secondary importance only.

4.5.3 Settlement and Related Risks

The Pfandbrief holder is subject to all kinds of risk that arise under financial transactions, such as *settlement risk*, fraud, etc. With respect to the importance of these risks in the context of German Pfandbriefe, the same argumentation as for operational risk holds.

4.5.4 Reputational Risk

Pfandbriefe are a high quality product with a long track record and no default until today. One single Pfandbrief default could, therefore, have a severe impact on the reputation of the whole Pfandbrief segment. In this context, systemic support (see Sect. 4.6.5) plays an important role as risk mitigant.

4.5.5 Regulatory Risk

The risk of (unexpected) changes to the regulatory framework is referred to as *regulatory risk*. As Pfandbrief holder protection is one of the main purposes of the PfandBG, it can be assumed that changes to this law would not be to the Pfandbrief holder's disadvantage. It can, however, not be precluded that changes to other parts of the regulatory framework have a negative impact on the Pfandbrief holder's situation.

A recent example in this context is the introduction of bank resolution systems in the European Union, the impact of which on the rights of Pfandbrief holders (and, more generally, covered bond holders) has been a major point of discussion recently, see for example ECBC (2011, 2012). In June 2012, the European Commission published its proposed EU directive on a framework for recovery and resolution of credit institutions (see European Commission 2012), which aims at increasing financial stability and minimizing taxpayer exposure to losses in case of bank insolvencies. The directive includes, among others, the so-called *bail-in tool*, which allows—under certain conditions—for a write-down of liabilities (i.e. a reduction of the creditors' legal claim), or a conversion of debt claims to equity. In the proposal, there seems to be no general exemption clause for covered bonds: Article 38 excludes secured liabilities from write-down and conversion powers, except for the case where the secured liabilities exceed the value of the pledged assets. In the context of covered bonds this means that, if the value of the cover pool is not sufficient to cover all outstanding covered bonds, the remaining senior unsecured claim (which ranks pari passu with the issuer's unsecured creditors) can be written down/bailed-in (Volk and Will 2012). However, Article 38 also states that member states can make certain exemptions under national law. The EU directive is planned to be finalized in 2013 and to come into force in 2015 (Moody's 2012a). At the time of writing, the directive and its overall implications were still being discussed. The German Restructuring Act ("Restrukturierungsgesetz"), see Bundesgesetzblatt (2010), which currently regulates the reorganisation and the orderly wind-up of credit institutions in Germany, already came into force in 2011. It excludes Pfandbriefe from direct bail-in. To make sure that Pfandbrief holders are protected from measures taken according to this law, the Pfandbrief Act was amended accordingly, through the introduction of a new paragraph, § 36a PfandBG. According to Volk and Will (2012), a similar preferential treatment is to be expected for the upcoming EU bank resolution regime. This does, however, not mean that Pfandbriefe will not at all be impacted by potential resolution measures. The bail-in of senior unsecured debt in the cover pool, for example, might have a negative impact on Pfandbrief holders (Moody's 2012a). Asset encumbrance and the availability of voluntary overcollateralization may also play a role in this context (for more details, see Sect. 4.6.2).

Other current examples of regulatory changes are Basel III (the new supervisor rules for banks) and Solvency II (the new regime for insurance companies). For more details on bail-in, Basel III and Solvency II and their implications for covered bond holders, see for example, ECBC (2011, 2012). As these examples show, regulatory risk is present and cannot be completely neglected. It is, however, hard to quantify. Rating agencies like Fitch explicitly exclude the risk of changes in covered bond legislation, which they refer to as *event risk*, when assigning a rating (Fitch 2012c). All in all, the above bail-in example still provides comfort that, for German Pfandbriefe, there is a comparably low level of regulatory risk as compared to other

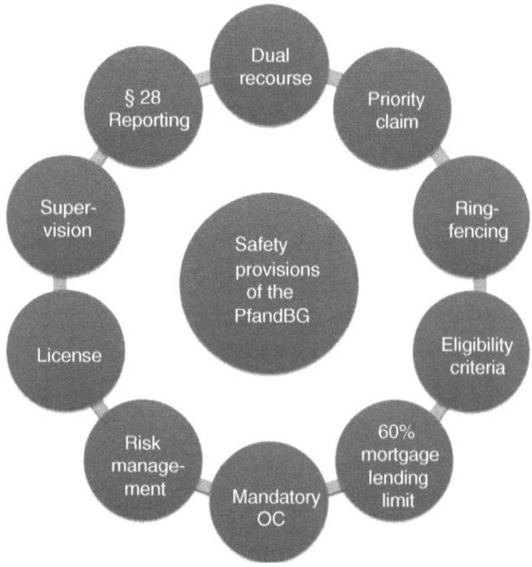

Fig. 28 The safety provisions of the PfandBG

financial products, which is mainly due the strong systemic support this product receives.

4.5.6 Conclusion

Sovereign risk has become more and more important over the last years. It impacts the Pfandbrief holder through issuer credit strength, cover pool credit quality, refinancing options and the market's perception of the likelihood of active state support. Regulatory risk is present and cannot be completely neglected. The other risks discussed in this section should be of secondary importance only.

4.6 Risk Mitigation

As already outlined in the above sections, Pfandbrief-related risks are mitigated by the safety provisions of the PfandBG and by program-specific features. Apart from that, supervision, transparency, systemic support and flexibility of legislation also play an important role.

4.6.1 Safety Provisions of the PfandBG

The safety provisions of the PfandBG as summarized by Fig. 28 ensure a high level of Pfandbrief quality. A detailed description of the legal framework for

Pfandbrief issuance was given in Sect. 2, while the role of the individual provisions was explained in the previous sections.

4.6.2 Program-Specific Features

Program-specific features that mitigate Pfandbrief risks are, for example, hedging activities, voluntary overcollateralization, and, more generally, the issuer's commitment to his Pfandbrief program.

Hedging Derivatives

Cover pool derivatives cannot only be used to hedge interest rate and currency risks, but also to amend cash flow structures, to close liquidity gaps, or to lock in interest rates for future covered bond issuances (Schulz 2011). When used appropriately, they can be a powerful tool to reduce cover pool risks. However, even if a cover pool is perfectly hedged against interest rate and currency risk, this can change over time as the cover pool is not static. Maturing, defaulting or prepaying assets may cause the cover pool to become over- or underhedged.

Hedging derivatives are themselves subject to risks, namely counterparty credit risk (see Sect. 4.2.1), and uncertainty regarding their continuation upon issuer default (see Sect. 4.4.1). In addition to that, the use of hedging derivatives is considered to add additional complexity to the cover pool, especially in a post-issuer insolvency scenario (Fitch 2012a). Rating agencies, whose main focus is on payment disruptions upon issuer insolvency, consider cover pool derivatives more and more critically, and in certain cases, hedging derivatives may even have a negative rating impact (Schulz 2012). As cover pool derivatives are currently not frequently used in German Pfandbrief programs, the whole subject is, however, of lower relevance. For a more detailed discussion on derivatives in the cover pool, see Schulz (2012) and the corresponding rating agency publications (Fitch 2012b; S&P 2012d; Moody's 2012c).

Voluntary Overcollateralization

In practice, many Pfandbrief issuers maintain a higher level of collateralization than the legally required one, as this is often needed to obtain a certain target rating. This overcollateralization is referred to as voluntary overcollateralization and provides an important cushion against all kinds of risk arising upon issuer insolvency, especially in the context of the cover pool. Voluntary overcollateralization is, however, itself subject to risk, as its maintenance is at the discretion of the issuer (see Sect. 4.1.2), and there are remaining uncertainties regarding its enforceability (see Sect. 4.4.1). The maintenance of voluntary overcollateralization also depends on the issuers commitment to the Pfandbrief program, as discussed in the following.

Issuer Commitment to His Pfandbrief Program

An issuer's commitment to his Pfandbrief program may go beyond legal obligations and provide additional ease to the Pfandbrief holder, especially in the context of issuer-related risks (see Sect. 4.1). Take, for example, the assurance of a certain level of voluntary overcollateralization (by contractual commitment or public statement), the maintenance of a certain target rating, or a committed business strategy. In the past, there was a large amount of such discretionary support from covered bond issuers towards their programs (Moody's 2013a). However, there is an increasing trend of covered bond (and also Pfandbrief) issuers to accept a rating downgrade instead of fulfilling the increasing overcollateralization requirements from rating agencies for AAA ratings, which is mainly caused by the costs associated with such overcollateralization levels (Bettink 2012). Fitch (2012c) assumes that the likelihood of reduced issuer commitment is especially high in case of wind-down or dormant programs. As already mentioned previously, it should, therefore, not be assumed that a current level of voluntary overcollateralization will be fully available at the time of issuer insolvency.

4.6.3 Supervision

In Germany, the regulator plays an active part in the supervision of Pfandbrief banks. There are regular cover pool audits, and an independent controlling body (the cover pool monitor) who reports his findings to BaFin. On top of the general banking supervision and the provisions of the PfandBG, BaFin has set up special divisions to supervise Pfandbrief issuers (see vdp 2013g):

- Pfandbrief Competence Center I (Fundamental Issues and Supervision of Pfandbrief Banks without Company Involvement), which focuses on legal and fundamental questions regarding the Pfandbrief Act, and
- Pfandbrief Competence Center II (Cover Audits of Pfandbrief Banks), which is responsible for regular cover pool audits that normally take place once every 2 years.

Fitch (2009c) considers the proactive monitoring of Pfandbrief banks by the German regulator as exemplary. In addition to that, there are also the rating agencies which constantly observe the Pfandbrief programs and, especially, the cover pools' quality.

4.6.4 Transparency

Transparency plays a crucial role and is especially important in the context of cover pool risks and substitution risk. The quarterly reporting requirements according to § 28 PfandBG as outlined in Sect. 2.9 bring transparency to the Pfandbrief market and ensure comparability (through the uniform format) and a certain frequency and scope of the reports. Over the last years, transparency has further increased through the amendments of the PfandBG and the vdp's transparency initiatives regarding § 28 reporting[37] and secondary market spread disclosure.[38] The information published according to § 28 PfandBG is, however, only available to the investor with a certain time lag and, even though transparency requirements are higher than in other countries, the availability of cover pool information still seems limited when compared to disclosure requirements in the securitization market. The FSA (the British supervisory authority), for example, requires covered bond issuers to disclose data on a loan level basis (Tolckmitt 2012a). In a recent position paper, the German Insurance Association (GDV) demands increased transparency beyond the latest amendments to § 28 PfandBG: more frequent (i.e. monthly) cover pool reporting and detailed information on residential and commercial cover pool assets and their origin (see GDV 2013). Dübel (2010) also demands higher transparency standards and a more frequent reporting.

4.6.5 Systemic Support

The systemic relevance of the German Pfandbrief as an important funding tool provides it with a high level of government support, which is considered exemplary by Fitch (2009c). In the past, the German Government has shown its support to the Pfandbrief by several Pfandbrief issuer bailouts, such as in the case of Allgemeine Hypothekenbank Rheinboden AG (October 2005), Düsseldorfer Hypothekenbank (April 2008), Hypo Real Estate AG (October 2008), Eurohypo AG (May 2009), see Kiff et al. (2011), and in 2008 it declared its unrestricted endorsement of the Pfandbrief (vdp 2008a):

> [T]he Pfandbrief is already secure—there has never been a case of default over its entire history of more than 200 years. The German Government shall ensure that this will also remain to be the case in future. If and when necessary for the proper functioning of the Pfandbrief market, the German Government shall therefore adopt further statutory measures on an ad-hoc basis in order to safeguard the German Pfandbrief.

Government support is not only restricted to bailout measures: it can also show in the form of assistance upon issuer default, such as providing liquidity support to guarantee timely payments to Pfandbrief holders, or ensuring an orderly transfer of the cover pool to another Pfandbrief bank (Moody's 2012c).

The sovereign debt crisis in Europe has illustrated that sovereign strength does not only impact the credit strength of the issuer and the quality of the cover pool, but also

---

[37] The vdp's § 28 reporting transparency initiative aligned reporting according to § 28 PfandBG across vdp member institutions by ensuring a uniform understanding of the legal requirements and a consistent format of the corresponding reports, which are available on the vdp's website. For more details see vdp Transparency initiative and Winkler (2010).

[38] To increase transparency on the secondary market for Pfandbriefe, the vdp started in January 2012 to publish daily average spreads for German Jumbo Pfandbriefe with a residual maturity of at least 2 years (vdp 2012b). Due to changed market requirements, in January 2013 the disclosed spreads were extended to include German Pfandbrief benchmark issuances (i.e. issuances with a minimum size of EUR 500 mn) and with a residual maturity of at least 1 year (vdp 2013a).

influences covered bond spreads through the market's perception of the likelihood of active state support. In this context, German Pfandbriefe clearly benefit from the German government's credit strength. There is, however, no basis for a claim on such implicit government guarantees (Koppmann 2009, p. 528), meaning that the Pfandbrief investor cannot take government intervention for granted. With the introduction of bank resolution systems in many countries (see Sect. 4.5.5), government support for failing banks is, in general, expected to decrease (Berninger and Winkler 2012). This also applies to German banks (Moody's 2011e). However, as pointed out by Moody's (2011b), reduced issuer support does not necessarily mean less support to the Pfandbrief product itself. Fitch (2012f) also assesses government support to remain strong in the case of German Pfandbriefe, but questions whether the support of the product will still continue to extend to the issuer given current developments in the context of bank resolution regimes.

In a well-established covered bond market as the German one, covered bonds also benefit from the support of other issuers in the same market which have an interest in safeguarding the product's reputation (Volk 2006, p. 153). Furthermore, on a European level there is support to the whole covered bond segment, which is, for example, shown by the 2009 and the 2011 covered bond purchase programs of the European Central Bank.

4.6.6 Flexibility of Legislation

The German Pfandbrief legislation has proved to be very flexible to quickly react to changed market environments. The 2009 amendment, for example, improved the German Pfandbrief's quality by the introduction of the liquidity buffer requirement, legal clarifications and increased transparency requirements (Stöcker 2009). One year later, the 2010 amendment strengthened the Pfandbrief by addressing the legal status of the cover pool and the cover pool administrator's possibilities to raise liquidity (Stöcker 2010). These amendments, which also included further changes not mentioned, were a reaction to changed market requirements, with the financial crisis causing refinancing risk and issuer solvency becoming more relevant. The same applies to the latest amendment of the PfandBG, which revises the rules concerning the cover pool administrator and further increases transparency requirements. This flexibility of the German Pfandbrief legislation contributes to the high quality of the Pfandbrief.

5 Resume

Since the recent financial crisis and the outbreak of the eurozone government debt crisis investors classify covered bonds not any more as a pure interest rate product. They now also consider the quality of the issuer, the legal basis for issuance, and the quality and type of the cover pool. The issuer's domicile country and systemic support to the issuer and the product have also become increasingly important. This

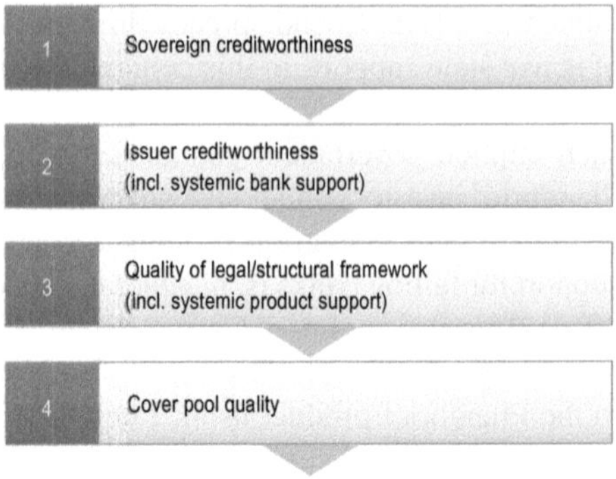

Fig. 29 Ranking of covered bond risk factors

has led to a differentiation of covered bond spreads—not only across jurisdictions but also within German Pfandbriefe. Liquidity is not any more the sole driver of Pfandbrief spreads: investors also require an additional risk premium for issuer risk and cover pool risk.

Across jurisdictions, sovereign risk and Europe's sovereign debt crisis have been the main drivers of covered bond spreads since 2010, even before issuer strength, the strength of the covered bond's legal/contractual framework, and the quality of the cover pool. Figure 29 illustrates the ranking of covered bond risk factors according to Packmohr (2011).

German Pfandbriefe have performed well during the financial crisis and the eurozone government debt crisis, and their spreads are at the lower end of the covered bond spectrum. The strong legislative framework with detailed provisions regarding asset quality, cover requirements and risk mitigation plays an important role in this context. The high level of systemic support, the strength of the German government, as well as supervision, transparency requirements and flexibility of legislation, also contribute to the Pfandbrief's benchmark status. Regarding the quality of Mortgage Pfandbriefe, the stability of the German real estate market plays an important role, while German public sector cover pools clearly benefit from the creditworthiness of the German sovereign. Should these prerequisites change, the high concentrations of German cover pool assets might, however, become an issue.

In the recent past, it has become more and more difficult for issuers to maintain top Pfandbrief ratings, mainly caused by a decrease in issuer ratings and increased rating agency requirements. As several Pfandbrief bank bailouts have shown, the default of a Pfandbrief bank is a potential risk that cannot be neglected. Moreover, like other covered bond laws, the PfandBG only provides protection against individual cover pool risks, but it cannot rule out the impact of systemic risks such as a collapse of the (local) financial system or a general deterioration of cover pool asset quality. In times of market distress, cover pool risks (notably refinancing risk, interest rate risk, currency risk and asset credit risk) can become a big issue. In addition to that, the investor is exposed to some remaining uncertainty caused by a lack of explicit legal provisions with respect to the protection of voluntary overcollateralization and the declaration of cover pool insolvency. Despite their high level of quality German Pfandbriefe can, therefore, not be considered to be risk-free.

References

Anhamm C (2010) Covered bond creditors vs. unsecured creditors. Presentation, October 2010

Avesani RG, Garca Pascual A, Ribakova E (2007) The Use of Mortgage Covered Bonds. IMF Working Paper No. 07/20, Feb

Bank of America Merrill Lynch (2010) Covered bond primer for the uninitiated. Research Report, 22.01.2010

Basel Committee on Banking Supervision (2006) International convergence of capital measurement and capital standards. Revised framework comprehensive version, June 2006

Baumgartner R (2012) The mortgage lending value: benefit or burden for Pfandbrief banks? In: Verband Deutscher Pfandbriebanken: The Pfandbrief 2012/2013: facts and figures about Europe's covered bond benchmark, Pfandbrief Fact Book, 17th edn. Berlin, pp 38–43

Berninger R (2010) Changes to the regulatory environment of Pfandbrief banks—beginning of a new era? In: Verband Deutscher Pfandbriefbanken: The Pfandbrief 2010/2011: facts and figures about Europe's covered bond benchmark, Pfandbrief Fact Book, 15th edn. Berlin, pp 64–75

Berninger R, Winkler S (2012) Asset encumbrance. In: European Covered Bond Council: European Covered Bond Fact Book, 7th edn, pp 88–102

Bollmus B (2006) Beleihungswertermittlung als Qualitätsgarantie für die Pfandbriefdeckung. Presentation, November 2006

Börsen-Zeitung (2013) Wir dürfen nicht alles über einen Kamm scheren. 19.01.2013

Bundesgesetzblatt (2005) Gesetz zur Neuordnung des Pfandbriefrechts vom 22. Mai 2005. Part I, No. 29. Bonn, 27.05.2005

Bundesgesetzblatt (2010) Gesetz zur Restrukturierung und geordneten Abwicklung von Kreditinstituten zur Errichtung eines Restrukturierungsfonds für Kreditinstitute und zur Verlängerung der Verjährungsfrist der aktienrechtlichen Organhaftung (Restrukturierungsgesetz). Part I, No. 63. Bonn, 14.10.2010

Bundesrat (2010) Gesetzentwurf der Bundesregierung: Entwurf eines Gesetzes zur Umsetzung der geänderten Bankenrichtlinie und der geänderten Kapitaladäquanzrichtlinie. Public Reference 155/10, 26.03.2010

Bundesrat (2013) Gesetzesbeschluss des Deutschen Bundestages: Gesetz zur Umsetzung der Richtlinie 2013/.../EU über den Zugang zur Tätigkeit von Kreditinstituten und die Beaufsichtigung von Kreditinstituten und Wertpapierfirmen und zur Anpassung des Aufsichtsrechts an die Verordnung (EU) Nr. .../2013 über die Aufsichtsanforderungen an Kreditinstitute und Wertpapierfirmen (CRD IV-Umsetzungsgesetz). Public Reference 374/13, 17.05.2013

Caris A (2012) Sub-jumbo sector. In: European Covered Bond Council: European Covered Bond Fact Book, 7th edn, pp 103–109

Commerzbank (2011a) Pfandbrief weekly. Covered bonds research, 13.04.2011

Commerzbank (2011b) Pfandbrief weekly. Covered bonds research, 27.10.2011

Commerzbank (2011c) Pfandbrief weekly. Covered bonds research, 23.11.2011

Crimmann W, Rüchardt K (2008) Mortgage lending value. vdp series of papers (40), Berlin

Deutsche Bundesbank (2011) Gesetz über das Kreditwesen - KWG, June 2011

Deutscher Bundestag (2009a) Antwort der Bundesregierung auf die Kleine Anfrage der Abgeordneten Dr. Volker Wissing, Jens Ackermann, Dr. Karl Addicks, weiterer Abgeordneter und der Fraktion der FDP - Drucksache 16/12934 - Gefährdung des Pfandbriefmarktes. Public Reference 16/13161, 26.05.2009

Deutscher Bundestag (2009b) Antwort der Bundesregierung auf die Kleine Anfrage der Abgeordneten Leo Dautzenberg, Otto Bernhardt, Klaus-Peter Flosbach, weiterer Abgeordneter und der Fraktion der CDU/CSU sowie der Abgeordneten Bernd Scheelen, Ingrid Arndt-Brauer, Lothar Binding (Heidelberg), weiterer Abgeordneter und der Fraktion der SPD - Drucksache 16/13713 - Sicherheit der Pfandbriefe in der Insolvenz einer Pfandbriefbank. Public Reference 16/13823, 21.07.2009

Deutscher Bundestag (2009c) Beschlussempfehlung und Bericht des 2. Untersuchungsausschusses nach Artikel 44 des Grundgesetzes. Public Reference 16/14000, 18.09.2009

Deutscher Bundestag (2006) Antwort der Bundesregierung auf die Kleine Anfrage der Abgeordneten Frank Schäffler, Jens Ackermann, Dr. Karl Addicks, weiterer Abgeordneter und der Fraktion der FDP - Drucksache 16/2734 - Verhalten der Bundesanstalt für Finanzdienstleistungsaufsicht bezüglich der Allgemeinen HypothekenBank Rheinboden AG. Public Reference 16/2949, 13.10.2006

Deutscher Bundestag (2005) Beschlussempfehlung und Bericht des Finanzausschusses (7. Ausschuss) zu dem Gesetzentwurf der Bundesregierung - Drucksachen 15/4321, 15/4487 - Entwurf eines Gesetzes zur Neuordnung des Pfandbriefrechts. Public Reference 15/4878, 16.02.2005

Deutscher Bundestag (2008) Entwurf eines Gesetzes zur Fortentwicklung des Pfandbriefrechts. Public Reference 16/11130, 01.12.2008

Deutscher Bundestag (2004) Entwurf eines Gesetzes zur Neuordnung des Pfandbriefrechts. Public Reference 15/4321, 29.11.2004

Deutscher Bundestag (2010) Entwurf eines Gesetzes zur Umsetzung der geänderten Bankenrichtlinie und der geänderten Kapitaladäquanzrichtlinie. Public Reference 17/1720, 17.05.2010

Deutscher Bundestag (2012) Gesetzentwurf der Bundesregierung: Entwurf eines Gesetzes zur Umsetzung der Richtlinie 2012/.../EU über den Zugang zur Tätigkeit von Kreditinstituten und die Beaufsichtigung von Kreditinstituten und Wertpapierfirmen und zur Anpassung des Aufsichtsrechts an die Verordnung (EU) Nr. .../2012 über die Aufsichtsanforderungen an Kreditinstitute und Wertpapierfirmen (CRD IV-Umsetzungsgesetz). Public Reference 17/10974, 15.10.2012

DG Hyp (2008) Der deutsche Pfandbriefmarkt 2008 | 2009. Research Publication, September 2008

DG Hyp (2012) The German Pfandbrief market 2012 | 2013. Research Publication, August 2012

Die Welt (2012) Bankenverband rettet einstige KarstadtQuelle-Bank. 30.05.2012. http://www.welt.de/print/die_welt/wirtschaft/article106390235/Bankenverband-rettet-einstige-KarstadtQuelle-Bank.html (last visited on 28.03.2013)

Dierks LH, Engelhard F (2010) Sind Covered Bonds sicherer als Staatsanleihen? In: Schiffmacher C (ed) Bond Yearbook (2010) - Das Nachschlagewerk für Emittenten und Investoren. Institutional Investment Verlag, 2010, pp 62–63

Dresch R (2010) Cover-specific structures and processes of a Pfandbrief bank. In: Verband Deutscher Pfandbriefbanken: The Pfandbrief 2010/2011: facts and figures about Europe's covered bond benchmark, Pfandbrief Fact Book, 15th edn. Berlin, 2010, pp 40–51

Dübel H-J (2010) Reformerfordernisse bei Pfandbriefen und Pfandbriefbanken vor dem Hintergrund der Finanzmarktkrise. Finpolconsult, Final Report, 21.06.2010

Eichert F (2012) In: European Covered Bond Council: European Covered Bond Fact Book, 7th edn, Chap. 5.1

Eichert F, Engelhard F (2012) Empirical evidence for the use of covered bonds under EU liquidity coverage rules. In: European Covered Bond Council: European Covered Bond Fact Book, 7th edn, pp 53–76

Engelhard F (2010) Liquidity management and asset disposal strategies in Covered bond issuer's stress scenarios. Presentation, November 2010

European Central Bank (2008) Covered bonds in the EU financial system. ECB Publication, December 2008

European Central Bank (2011) The impact of the Eurosystem's covered bond purchase programme on the primary and secondary markets. Occasional Paper Series No 122, January 2011

European Central Bank (2013) Eurosystem credit assessment framework (ECAF). http://www.ecb.int/paym/coll/elisss/ratingscale/html/index.en.html (last visited on 28.03.2013)

European Commission (2012) Proposal for a Directive of the European Parliament and of the Council establishing a framework for the recovery and resolution of credit institutions and investment firms and amending Council Directives 77/91/EEC and 82/891/EC, Directives 2001/24/EC, 2002/47/EC, 2004/25/EC, 2005/56/EC, 2007/36/EC and 2011/35/EC and Regulation (EU) No 1093/2010. Brussels, 06.06.2012

European Commission (2013) The Euro. http://ec.europa.eu/economy_finance/euro/index_en.htm (last visited on 28.03.2013)

European Covered Bond Council (2008) European Covered Bond Fact Book, 3rd edn

European Covered Bond Council (2009) European Covered Bond Fact Book, 4th edn
European Covered Bond Council (2010) European Covered Bond Fact Book, 5th edn
European Covered Bond Council (2011) European Covered Bond Fact Book, 6th edn
European Covered Bond Council (2012) Covered bond statistics. In: European Covered Bond Council: European Covered Bond Fact Book, 7th edn., Chap. 5
European Covered Bond Council (2012) European Covered Bond Fact Book, 7th edn
European Covered Bond Council. ECBC Essential Features of Covered Bonds
FitchRatings (2009a) Comparative study of covered bonds 2008/09. Covered bonds global special report, 06.05.2009
FitchRatings (2009b) Contractual mechanisms in covered bonds: under the spotlight. Covered bonds global special report, 04.06.2009
FitchRatings (2009c) Deutsche Postbank AG: public sector Pfandbriefe. New issue report, 20.07.2009
FitchRatings (2009d) Revised market value decline assumptions for German residential mortgage-backed transactions. Structured finance criteria report, 09.06.2009
FitchRatings (2010a) Application of the refinancing register in German structured finance transactions. Structured finance Germany cross-sector criteria, 15.12.2010
FitchRatings (2010b) Assessment of liquidity risks in covered bonds. Covered bonds global cross-sector criteria, 16.08.2010
FitchRatings (2011a) 2012 Outlook: Covered Bonds. Outlook Report, 22.12.2011
FitchRatings (2011b) Banks' use of covered bond funding on the rise. Covered bonds global special report, 10.03.2011
FitchRatings (2011c) Covered bonds investor survey. Credit market research, EMEA Special Report, 07.02.2011
FitchRatings (2011d) EMEA criteria addendum—Germany. Structured finance sector-specific criteria report, 10.08.2011
FitchRatings (2011e) Trends in bank funding profiles: secured financing on the rise, but likely to tail off. Banks global special report, 16.06.2011
FitchRatings (2012a) 2013 Outlook: Covered Bonds. Outlook Report, 17.12.2012
FitchRatings (2012b) Covered bonds counterparty criteria. Covered bonds global cross-sector criteria, 25.07.2012
FitchRatings (2012c) Covered bonds rating criteria. Master criteria, 18.09.2012
FitchRatings (2012d) Covered bonds rating criteria—mortgage liquidity & refinance stress addendum. Sector-specific rating criteria, 14.11.2012
FitchRatings (2012e) EMEA Residential mortgage loss criteria. Sector-specific criteria, 07.06.2012
FitchRatings (2012f) Support is key to German Pfandbrief issuers' ratings. Press release, 28.08.2012. http://www.fitchratings.com/creditdesk/press_releases/detail.cfm?pr_id=758902 (last visited on 28.03.2013)
FitchRatings (2013a) Asset analysis criteria for covered bonds of European public entities. Sector-specific rating criteria, 30.01.2013
FitchRatings (2013b) Counterparty criteria for structured finance and covered bonds. Structured finance global cross-sector criteria, 13.05.2013
FitchRatings (2013c) Counterparty criteria for structured finance and covered bonds: derivative addendum. Structured finance global cross-sector criteria, 13.05.2013
FitchRatings (2013d) Covered bonds rating criteria—public sector liquidity and spread assumption addendum. Sector-specific rating criteria, 01.02.2013
FitchRatings (2013e) Criteria for the analysis of commercial real estate loans securing covered bonds. Sector-specific rating criteria, 07.02.2013
FitchRatings (2004) Analysis of German Pfandbriefe: Preparing for 2005. Special Report, 27.10.2004
FitchRatings. The impact of covered bonds issuance on banks ratings. Covered bonds global special report, 29.09.2008

FitchTraining (2010) Covered bonds credit and market risk workshop. Course materials, November 2010
Forster Y, Heberlein HM, Sirotic A (2011) Covered bond ratings. In: Pinedo AT, Tanenbaum JR (ed) Covered Bonds Handbook (Chap. 7), vol 1. Practising Law Institute, New York
Fragen an Jan Bettink zur Zukunft des Pfandbriefs. In: Immobilien & Finanzierung - Der Langfristige Kredit 14, 15.07.2012, p 469
FTSE Global Markets (2010) The financial crisis: stress-testing the Pfandbrief. Pfandbrief Roundtable, June 2010
Gesamtverband der Deutschen Versicherungswirtschaft (2013) Stellungnahme des Gesamtverbandes der Deutschen Versicherungswirtschaft zum Entwurf des CRD IV Umsetzungsgesetzes (Drucksache 17/10974). Berlin, 02.05.2013
Golin J (ed) (2006) Covered bonds beyond Pfandbriefe: innovations investment and structured alternatives. Euromoney Institutional Investor Plc, London
Grossmann R, Stöcker O (2012) Overview of covered bonds. In: European Covered Bond Council European Covered Bond Fact Book, 7th edn., pp 149–159
Hillenbrand F (2011) Timely payment in issuer's insolvency. Presentation, November 2011
Holter R (2006) Inhalt und Wirkung der BelWertV, In: Immobilien & Finanzierung 14–2006, pp 478–480
International Swaps And Derivatives Association (2010) ISDA margin survey 2010
Kiff J, Surti J, Jobst A (2011) Covered bonds and asset encumbrance. In: European Covered Bond Council: European Covered Bond Fact Book, 6th edn., pp 79–82
Koppmann T (2009) Gedeckte Schuldverschreibungen in Deutschland und Großbritanien. Schriften zum Europäischen und Internationalen Privat-, Bank- und Wirtschaftsrecht (31). De Gruyter Rechtswissenschaften Verlags-GmbH, Berlin
Kullmann W (2011) Covered bond case study—Germany. In: Pinedo AT, Tanenbaum JR (ed) Covered Bonds Handbook (Chap. 3), vol 1. Practising Law Institute, New York
Labe M (2010) The cover pool monitor of a Pfandbrief bank: duties—powers—limits. In: Verband Deutscher Pfandbriefbanken: The Pfandbrief 2010/2011: facts and figures about Europe's covered bond benchmark, Pfandbrief Fact Book, 15th edn. Berlin, 2010, pp 34–39
Melms M (2012) A comparison of rating approaches for Pfandbriefe. In: Verband Deutscher Pfandbriefbanken: The Pfandbrief 2012/2013: facts and figures about Europe's covered bond benchmark, Pfandbrief Fact Book, 17th edn. Berlin, 2012, pp 64–71
Moody's Investors Service (2008a) 2007 Review & 2008 Outlook. Special report 01.02.2008
Moody's Investors Service (2008b) Timely payment in covered bonds following sponsor bank default. Rating methodology, 13.03.2008
Moody's Investors Service (2009a) Annual sector review: covered bonds. Special report, 29.06.2009
Moody's Investors Service (2009b) Assessing swaps as hedges in the covered bond market. Rating methodology, 01.12.2009
Moody's Investors Service (2009c) EMEA covered bonds: 2008 Review & 2009 Outlook. Special report 26.01.2009
Moody's Investors Service (2009d) Moody's updated MILAN methodology for rating German RMBS. Rating methodology, 29.09.2009
Moody's Investors Service (2010a) A short guide to covered bonds. Structured finance in focus, 2010
Moody's Investors Service (2010b) Amendments to the Pfandbrief act would improve credit quality of German covered bonds. Sector comment, March 2010
Moody's Investors Service (2010c) EMEA covered bonds: 2010 Outlook & 2009 Review. Special report 04.02.2010
Moody's Investors Service (2010d) Moody's analysis of legal framework for German Pfandbrief. Special report, 01.07.2010
Moody's Investors Service (2010e) Moody's EMEA covered bond monitoring overview: Q3 2009. Special report, 13.04.2010

Moody's Investors Service (2010f) Moody's EMEA covered bonds monitoring overview: Q4 2009. Special report, 21.07.2010
Moody's Investors Service (2010g) Moody's EMEA covered bonds monitoring overview: Q1 2010. Special report, 21.09.2010
Moody's Investors Service (2010h) Moody's EMEA covered bonds monitoring overview: Q2 2010. Special report, 24.12.2010
Moody's Investors Service (2011a) EMEA covered bonds: 2011 outlook and 2010 review. Special report 17.01.2011
Moody's Investors Service (2011b) European covered bonds: 2012 outlook. Special report 12.12.2011
Moody's Investors Service (2011c) Moody's EMEA covered bonds monitoring overview: Q3 2010. Special report, 14.04.2011
Moody's Investors Service (2011d) Moody's EMEA covered bonds monitoring overview: Q1 2011. Special report, 08.07.2011
Moody's Investors Service (2011e) Moody's reduces support assumptions for German landesbanken. Special comment, 16.11.2011
Moody's Investors Service (2011f) Moody's EMEA covered bonds monitoring overview: Q2 2011. Special report, 07.10.2011
Moody's Investors Service (2011g) The great credit shift: structured finance credit post crisis. Report, 12.09.2011
Moody's Investors Service (2012a) Global covered bonds: 2013 outlook. Special comment 13.12.2012
Moody's Investors Service (2012b) How sovereign credit quality may affect other ratings. Rating implementation guidance, 13.02.2012
Moody's Investors Service (2012c) Moody's approach to rating covered bonds. Rating methodology, 27.07.2012
Moody's Investors Service (2012d) Moody's European covered bonds monitoring overview: Q3 2011. Special report. 19.01.2012
Moody's Investors Service (2012e) Moody's European covered bonds monitoring overview: Q1 2012. Special report, 19.07.2012
Moody's Investors Service (2012f) Moody's European covered bonds monitoring overview: Q2 2012. Special comment, 09.10.2012
Moody's Investors Service (2012g) Moody's European covered bonds monitoring overview: Q4 2011. Special report, 19.04.2012
Moody's Investors Service (2012h) Moody's: asset selection risk increases for German public-sector Pfandbriefe. Announcement, 04.04.2012. http://www.moodys.com/research/Moodys-Asset-selection-risk-increases-for-German-public-sector-Pfandbriefe-PR_242613 (last visited on 28.03.2013)
Moody's Investors Service (2013a) Moody's European covered bonds monitoring overview: Q3 2012. Special comment, 28.01.2013
Moody's Investors Service (2013b) Moody's says European SME covered bond market expected to expand. Announcement, 05.06.2013. http://www.moodys.com/research/Moodys-says-European-SME-covered-bond-market-expected-to-expand-PR_274911 (last visited on 07.06.2013)
Moody's Investors Service (2005a) European covered bond legal frameworks: Moody's legal check-list. Special report, 09.12.2005
Moody's Investors Service (2005b) Moody's rating approach to European covered bonds. Rating methodology, 13.06.2005
Moser S, Pesaresi N, Soukup K (2002) State guarantees to German public banks: a new step in the enforcement of State aid discipline to financial services in the Community. In: Competition policy newsletter, Number 2, June 2002
NordLB (2011) Covered bond view. Fixed income research, 09.11.2011

Packer F, Stever R, Upper C (2007) The covered bond market. BIS quarterly review, September 2007

Packmohr T (2011) Risks involved in using public-sector debt as collateral for covered bonds. Presentation, November 2011

Pfandbriefe: Investors Darling. In: Immobilien & Finanzierung - Der Langfristige Kredit 09, 01.05.2013, p 314

Pinedo AT, Tanenbaum JR (eds) (2011a) Covered Bonds Handbook, vol 1. Practising Law Institute, New York

Pinedo AT, Tanenbaum JR (eds) (2011b) Covered Bonds Handbook, vol 2. Practising Law Institute, New York

Prokopczuk M, Vonhoff V (2012) Risk premia in covered bond markets. J Fixed Income 22(2):19–29

Prokopczuk M, Siewert JB, Vonhoff V (Forthcoming) Credit risk in covered bonds. J Empir Finan

Quentin J (2008) The subprime crisis—implications for property valuation? The revival of the mortgage lending value. In: Verband Deutscher Pfandbriefbanken (2008): Real estate banking 2008/2009: Real estate banking—committed to professionalism—facts and figures, Real Estate Fact Book, 7th edn. Berlin, pp 24–33

Rehse D, Medla H, Landgraf D. DunkleWolken (2010) am US-Markt für Gewerbeimmobilienfinanzierungen: Droht ein ähnliches Szenario in Deutschland? EBS Discussion Paper, Wiesbaden 15.06.2010

Rudolf F, Hillenbrand F, Langer H (2012) Timely payment and the role of soft-bullet structures. In: European Covered Bond Council: European Covered Bond Fact Book, 7th edn., pp 131–139

Rudolf M, Saunders A (2009) Refinancing real estate loans—lessons to be learned from the subprime crisis. vdp series of papers (38), Berlin

Sarafana J (2010) Covered bond performance and sovereign spreads. In: European Covered Bond Council: European Covered Bond Fact Book, 5th edn., pp 52–57

Schiffmacher C (ed) (2010) Bond yearbook 2010 - Das Nachschlagewerk für Emittenten und Investoren. Institutional Investment Verlag

Schönfeld D (2012) Knappes Gut Pfandbrief. In: Immobilien & Finanzierung - Der Langfristige Kredit 14, 15.07.2012, pp 22–23

Schulz M (2010) Covered bond legislations—a valuation model. Presentation, November 2010

Schulz M (2011) Derivatives in cover pools. Presentation, November 2011

Schulz M (2012) Derivatives in cover pools. In: European Covered Bond Council: European Covered Bond Fact Book, 7th edn., pp 125–130

Siewert JB, Vonhoff V (2011) Liquidity and credit risk premia in the Pfandbrief market. Working paper, September (2011)

Soldera J, Lindstrom N, Soriano JP (2012) Moody's covered bond rating method. In: European Covered Bond Council: European Covered Bond Fact Book, 7th edn., pp 535–540

Sommer T (2012) Die Zukunft des öffentlichen Pfandbriefs. In: Immobilien & Finanzierung - Der Langfristige Kredit 01, 02.01.2012, pp 19–21

Standard&Poor's (2004a) German Pfandbrief framework further improved. Commentary report, 15.04.2004

Standard&Poor's (2004b) Improvements to German Pfandbrief framework should bring greater clarity for investors. Commentary report, 22.01.2004

Standard&Poor's (2004c) Rating Pfandbriefe—the analytical perspective. Rating criteria, 08.04.2004

Standard&Poor's (2005a) Criteria update: refined commercial real estate analysis in European covered bonds. Commentary report, 18.11.2005

Standard&Poor's (2005b) Extended criteria on collateral agreements for swaps in the cover register of covered bonds. Commentary report, 05.01.2005

Standard&Poor's (2005c) Germany farewells mortgage bank act for new covered bond act to take center stage. Commentary report, 21.03.2005

Standard&Poor's (2007a) All covered bonds are not created equal. RatingsDirect, 13.09.2007

Standard&Poor's (2007b) German refinancing registers could help source assets for Pfandbriefe. RatingsDirect, 10.10.2007
Standard&Poor's (2007c) Overcollateralization adequately protecting covered bond ratings, despite market conditions. Structured Finance, 08.11.2007
Standard&Poor's (2007d) Standard & Poor's Ratingansatz für Covered Bonds und Pfandbriefe. Structured finance, 2007
Standard&Poor's (2012a) Assessing asset-liability mismatch risk in covered bonds: revised methodology and assumptions for target asset spreads. Rating criteria, 24.04.2012
Standard&Poor's (2012b) Counterparty risk framework methodology and assumptions. Rating criteria, 29.11.2012
Standard&Poor's (2012c) Covered bond ratings framework: methodology and assumptions. Rating criteria, 26.06.2012
Standard&Poor's (2012d) Covered bonds counterparty and supporting obligations methodology and assumptions. RatingsDirect, 31.05.2012
Standard&Poor's (2012e) What factors do we consider when analyzing commingling and account bank risk in covered bonds? Credit FAQ, 26.11.2012
Standard&Poor's (2013) Covered bonds counterparty criteria have limited negative rating impact. RatingsDirect, 20.02.2013
Standard&Poor's (2011) Never underestimate credit risk in mortgage covered bonds. RatingsDirect, 12.09.2011
Standard&Poor's (2009) Revised methodology and assumptions for assessing asset-liability mismatch risk in covered bonds. RatingsDirect, 16.12.2009
Stöcker O (2005) New legislation on covered bonds in Europe: structure of issuer, ALM, banking supervision, segregation and bankruptcy remoteness. Presentation, Oct 2005
Stöcker O (2009) Amendment of the Pfandbrief act (2009) In: Verband Deutscher Pfandbriefbanken: The Pfandbrief 2009/2010: facts and figures about Europe's covered bond benchmark, Pfandbrief Fact Book, 14th edn. Berlin, 2009, pp 18–23
Stöcker O (2010) Amendment of the Pfandbrief Act. In: Verband Deutscher Pfandbriefbanken: The Pfandbrief 2010/2011: facts and figures about Europe's covered bond benchmark, Pfandbrief Fact Book, 15th edn. Berlin, 2010, pp 20–25
Stöcker O (2012) Asset encumbrance and German Pfandbriefe. In: Verband Deutscher Pfandbriefbanken: The Pfandbrief 2012/2013: facts and figures about Europe's covered bond benchmark, Pfandbrief Fact Book, 17th edn. Berlin, 2012, pp 32–37
Stürner R, Kern C (2007) Deutsche Hypothekenpfandbriefe und U.S.-amerikanische Deckungswerte. vdp series of papers (28), Berlin
Sünderhauf R (2006) Bewertung des Ausfallrisikos deutscher Hypothekenbank-Pfandbriefe. BWV Berliner Wissenschafts-Verlag GmbH, 2006
The Covered Bond Report (2012) The Pfandbrief roundtable 2012. January/February 2012
Tolckmitt J (2011) Der Pfandbrief - Qualitatsbenchmark am Covered Bond Markt. Presentation, June 2011
Tolckmitt J (2012a) Transparenz als Wettbewerbsvorteil - Ziele der PfandBG-Novelle 2012. In: Immobilien & Finanzierung - Der Langfristige Kredit 14, 15.07.2012, pp 470–471
Tolckmitt J, Stöcker O (2012b) The legal framework for issuing Pfandbriefe. In: Verband Deutscher Pfandbriefbanken: The Pfandbrief 2012/2013: facts and figures about Europe's covered bond benchmark, Pfandbrief Fact Book, 17th edn. Berlin, 2012, pp 6–21
Verband Deutscher Pfandbriefbanken The mortgage lending value—sustainability since 1900. Brochure
Verband Deutscher Pfandbriefbanken (2005a) Real estate banking: Real estate banking—committed to professionalism—facts and figures 2005, Real Estate Fact Book, 4th edn. Berlin
Verband Deutscher Pfandbriefbanken (2005b) The Pfandbrief—Europe's covered bond benchmark: facts and figures 2005. Pfandbrief Fact Book, 10th edn. Berlin
Verband Deutscher Pfandbriefbanken (2005c) Das Pfandbriefgesetz - Textsammlung und Materialien. Fritz Knapp Verlag GmbH, Frankfurt am Main

Verband Deutscher Pfandbriefbanken (2006a) Real estate banking 2006: Real estate banking—committed to professionalism—facts and figures, Real Estate Fact Book, 5th edn. Berlin
Verband Deutscher Pfandbriefbanken (2006b) The Pfandbrief 2006: facts and figures about Europe's covered bond benchmark. Pfandbrief Fact Book, 11th edn. Berlin
Verband Deutscher Pfandbriefbanken (2007a) Real estate banking 2007/2008: Real estate banking—committed to professionalism—facts and figures, Real Estate Fact Book, 6th edn. Berlin
Verband Deutscher Pfandbriefbanken (2007b) The Pfandbrief 2007/2008: facts and figures about Europe's covered bond benchmark. Pfandbrief Fact Book, 12th edn. Berlin
Verband Deutscher Pfandbriefbanken (2008a) German financial market stabilization act: unrestricted endorsement of the Pfandbrief. vdp press release no. 13, 13.10.2008
Verband Deutscher Pfandbriefbanken (2008b) Real estate banking 2008/2009: Real estate banking—committed to professionalism—facts and figures, Real Estate Fact Book, 7th edn. Berlin
Verband Deutscher Pfandbriefbanken (2008c) The Pfandbrief 2008/2009: facts and figures about Europe's covered bond benchmark. Pfandbrief Fact Book, 13th edn. Berlin
Verband Deutscher Pfandbriefbanken (2009a) Real estate banking 2009/2010: Real estate banking—committed to professionalism—facts and figures, Real Estate Fact Book, 8th edn. Berlin
Verband Deutscher Pfandbriefbanken (2009b) The Pfandbrief 2009/2010: facts and figures about Europe's covered bond benchmark. Pfandbrief Fact Book, 14th edn. Berlin
Verband Deutscher Pfandbriefbanken (2010a) Annual report 2009. Berlin
Verband Deutscher Pfandbriefbanken (2010b) Funding activities of the German Pfandbrief banks. In: Verband Deutscher Pfandbriefbanken: Annual report 2009, Berlin, pp 7–17
Verband Deutscher Pfandbriefbanken (2010c) Real estate banking 2010/2011: Real estate banking—committed toprofessionalism—facts and figures. Real Estate Fact Book, 9th edn. Berlin
Verband Deutscher Pfandbriefbanken (2010d) The Pfandbrief 2010/2011: facts and figures about Europe's covered bond benchmark. Pfandbrief Fact Book, 15th edn. Berlin
Verband Deutscher Pfandbriefbanken (2011a) Annual report 2010. Berlin
Verband Deutscher Pfandbriefbanken (2011b) Real estate banking 2011/2012: Real estate banking—committed to professionalism—facts and figures, Real Estate Fact Book, 10th edn. Berlin
Verband Deutscher Pfandbriefbanken (2011c) The Pfandbrief 2011/2012: facts and figures about Europe's covered bond benchmark. Pfandbrief Fact Book, 16th edn. Berlin
Verband Deutscher Pfandbriefbanken (2011d) Pfandbrief Act (as at January 2011). Brochure
Verband Deutscher Pfandbriefbanken (2012a) Annual report 2011. Berlin
Verband Deutscher Pfandbriefbanken (2012b) Greater transparency for Jumbo Pfandbriefe. vdp press release no. 1, 09.01.2012
Verband Deutscher Pfandbriefbanken (2012c) Real estate banking 2012/2013: Real estate banking—committed to professionalism—facts and figures, Real Estate Fact Book, 11th edn. Berlin
Verband Deutscher Pfandbriefbanken (2012d) The Pfandbrief 2012/2013: facts and figures about Europe's covered bond benchmark. Pfandbrief Fact Book, 17th edn. Berlin
Verband Deutscher Pfandbriefbanken (2012e) Volume outstanding and issuance 2003–2011. Pfandbrief market statistics
Verband Deutscher Pfandbriefbanken (2013a) Greater transparency on the secondary market for Pfandbriefe. vdp press release no. 1, 07.01.2013
Verband Deutscher Pfandbriefbanken (2013b) Jahresbericht 2012. Berlin
Verband Deutscher Pfandbriefbanken (2013c) Jumbo Pfandbriefe. http://www.pfandbrief.de/cms/_internet.nsf/tindex/en_131.htm (last visited on 10.04.2013)
Verband Deutscher Pfandbriefbanken (2013d) Pfandbrief issues in 2012: Mortgage Pfandbrief continues to rise in importance. vdp press release no. 4, 18.02.2013

Verband Deutscher Pfandbriefbanken (2013e) Pfandbrief statistics pursuant to § 28 Pfandbrief Act. http://www.pfandbrief.de/cms/_internet.nsf/tindex/en_pub_pfandbg.htm (last visited on 28.03.2013)
Verband Deutscher Pfandbriefbanken (2013f) Spotlight real estate—German housing market at the start of 2013. Publication, Berlin, January 2013
Verband Deutscher Pfandbriefbanken (2013g) Supervision of Pfandbrief banks. http://www.pfandbrief.de/cms/_internet.nsf/tindex/en_115.htm (last visited on 28.03.2013)
Verband Deutscher Pfandbriefbanken (2013h) Traditional Pfandbriefe. http://www.pfandbrief.de/cms/_internet.nsf/tindex/en_132.htm (last visited on 10.04.2013)
Verband Deutscher Pfandbriefbanken (2013i) vdp anticipates stable year on Pfandbrief market—Pfandbriefe attract new issuers and investors. vdp press release no. 5, 17.04.2013
Verband Deutscher Pfandbriefbanken (2013j) Volume outstanding and issuance 2003–2012. Pfandbrief market statistics
Verband Deutscher Pfandbriefbanken (2013k) http://www.pfandbrief.de (last visited on 28.03.2013)
Verband Deutscher Pfandbriefbanken. Transparency initiative. Brochure
Voigtländer M (2009) Volkswirtschaftliches Basiswissen Immobilien. vdp Publication, Berlin
Volk B (2006) Major covered bond markets: Germany. In: Golin J (ed) Covered bonds beyond Pfandbriefe: innovations, investment and structured alternatives, Euromoney Institutional Investor Plc, London, pp 134–159
Volk B (2009) Overview covered bonds: adjusting to the new market. Deutsche Bank, Global Markets Research, 04.02.2009
Volk B (2011a) Covered bond market—set for further structural changes. Deutsche Bank, Global Markets Research, 24.01.2011
Volk B (2011b) Covered bonds—heterogeneous market in structural change. Presentation, March 2011
Volk B (2011c) RMBS vs. covered bonds. In: European Covered Bond Council: European Covered Bond Fact Book, 6th edn., pp 109–116
Volk B (2012) The Pfandbrief market 2011/2012. In: Verband Deutscher Pfandbriefbanken: The Pfandbrief 2012/2013: facts and figures about Europe's covered bond benchmark, Pfandbrief Fact Book, 17th edn. Berlin, 2012, pp 22–31
Volk B, Will F (2012) Covered bonds versus senior unsecured bank debt and RMBS. In: European Covered Bond Council: European Covered Bond Fact Book, 7th edn., pp 206–222
Will F (2011) Covered bonds versus senior unsecured bank debt. In: European Covered Bond Council: European Covered Bond Fact Book, 6th edn., pp 61–71
Winkler B (2010) vdp transparency initiative. In: Verband Deutscher Pfandbriefbanken: The Pfandbrief 2010/2011: facts and figures about Europe's covered bond benchmark, Pfandbrief Fact Book, 15th edn. Berlin, 2010, pp 52–57
Wolf D (2012) Pfandbriefe in institutional asset management. In: Verband Deutscher Pfandbriefbanken: The Pfandbrief 2012/2013: facts and figures about Europe's covered bond benchmark, Pfandbrief Fact Book, 17th edn. Berlin, 2012, pp 44–51
Wolke T (2008) Risikomanagement, 2nd edn. Oldenbourg Wissenschaftsverlag GmbH, München
Zentraler Kreditausschuss (2008) Stellungnahme des Zentralen Kreditausschusses zum Referentenentwurf für die Novellierung des PfandBG. VII B 3 - WK 7100/08/10002-1, Berlin, 26.08.2008

Glossary

§ 28 transparency requirements According to § 28 PfandBG, Pfandbrief banks are required to publish key figures related to their Pfandbrief business on a quarterly basis.

Acceleration When a Pfandbrief is accelerated, it becomes due and payable immediately.

Aircraft Pfandbrief A Pfandbrief backed by mortgages on aircraft.

Amortization test Contractual feature that defines specific trigger events for cover pool insolvency, which result in cover pool acceleration in case of breach.

Anstaltslast German state guarantee scheme for public sector credit institutions (such as German savings banks and Landesbanks), which was replaced by a normal commercial owner relationship in July 2001, with an interims phase from July 2001 to 2005. (Source: Moser et al. 2002).

Asset credit risk Risk of losses arising from the cover pool due to changing credit quality of its assets, including asset default risk, downgrade risk and credit spread risk.

Asset encumbrance Reservation of certain assets on the bank's balance sheet for specific creditors, mainly through central bank funding, derivative transactions, covered bond issuance, other types of secured funding and securitization.

BaFin The German Financial Supervisory Authority ("Bundesanstalt für Finanzdienstleistungsaufsicht").

Bail-in tool Under the proposed EU directive on a framework for the recovery and resolution of credit institutions, the bail-in tool allows, under certain conditions, for a write-down of liabilities (i.e. a reduction of the creditors' legal claim), or a conversion of debt claims to equity.

BelWertV Regulation on the Determination of the Mortgage Lending Value ("Beleihungswertermittlungsverordnung").

M. Spangler and R. Werner, *German Covered Bonds*,
SpringerBriefs in Finance, DOI: 10.1007/978-3-319-02553-7,

Benchmark covered bond A covered bond denominated in euros, with a bullet maturity, fix annual coupons and a minimum outstanding volume of at least EUR 500 mn. There are also requirements for the syndicate banks to quote two-way prices. (Source: Grossmann and Stöcker 2012).

Bullet maturity The principal of a loan/bond with a bullet maturity is not repaid over its life but in a lump sum at maturity.

Bullet repayment See bullet maturity.

Claw-back risk Risk that the issuer's general insolvency administrator tries to claw back cover pool assets (i.e. to reverse their assignment to the cover pool).

Commingling risk Risk that cash flows belonging to the cover pool cannot be fully separated from cash flows belonging to the issuer's general insolvency estate (commingling credit risk), or only with delay (commingling liquidity risk).

Concentration risk Uneven distribution of credit exposure to particular borrowers or sectors.

Counterparty (credit) risk Risk of losses arising from the cover pool due to changing credit quality of its derivative counterparties.

Country risk Risk of losses due to a foreign sovereign's inability or unreadiness to provide the required foreign currency to make timely payments, including sovereign (credit) risk, transfer risk and political risk.

Cover pool Cover assets and derivatives which are registered in the cover register.

Cover pool administrator Upon issuer insolvency—or even earlier if deemed necessary by BaFin—at least one cover pool administrator ("Sachwalter") is appointed. His main task is managing the cover pool and ensuring timely payments to the privileged creditors until their claims are fully satisfied.

Cover pool insolvency Cover pool over-indebtedness or failure to pay.

Cover pool insolvency timing Risk of losses arising from the fact that the Pfandbrief Act does not specify specific criteria or trigger events regarding the declaration of cover pool insolvency.

Cover pool liquidation risk Risk of losses arising from the fact that the Pfandbrief Act does not define specific criteria or trigger events regarding the declaration of cover pool insolvency (cover pool insolvency timing), and that there are no restrictions regarding the amount of assets that can be liquidated at once for the repayment of a specific Pfandbrief maturity (time subordination).

Cover pool monitor At each Pfandbrief bank, an independent cover pool monitor ("Treuhänder") and at least one deputy is appointed. His rights and responsibilities are clearly defined by the Pfandbrief Act, and his main task is to safeguard the quality of the Pfandbrief.

Credit quality step	1	2	3
Moody's	Aaa, Aa1, Aa2, Aa3	A1, A2, A3	Baa1, Baa2, Baa3
Fitch	AAA, AA+, AA, AA-	A+, A, A-	BBB+, BBB, BBB-
S&P	AAA, AA+, AA, AA-	A+, A, A-	BBB+, BBB, BBB-

Fig. 1 Overview of credit quality steps (Source: ECB 2013)

Cover pool risks Risks related to the performance of the cover pool, including credit risk, interest rate risk, currency risk, refinancing risk, reinvestment risk, prepayment risk and real estate risk.

Cover register Cover pool assets and cover pool derivatives are made identifiable through registration in the cover register.

Covered bond There is no official definition for covered bonds, but the European Covered Bond Council has specified some essential features: A covered bond can be special-law based or general-law based, and must fulfil the following requirements (ECBC Essential Features of Covered Bonds):

> 1. The bond is issued by—or bondholders otherwise have full recourse to—a credit institution which is subject to public supervision and regulation; 2. Bondholders have a claim against a cover pool of financial assets in priority to the unsecured creditors of the credit institution; 3. The credit institution has the ongoing obligation to maintain sufficient assets in the cover pool to satisfy the claims of covered bondholders at all times; 4. The obligations of the credit institution in respect of the cover pool are supervised by public or other independent bodies.

Credit quality step The Eurosystem's harmonised rating scale distinguishes between three different credit quality steps, cf. Fig. 1.

Credit risk Risk of losses arising from the cover pool due to changing credit quality of its assets (asset credit risk) or its derivative counterparties (counterparty risk).

Credit spread risk Risk of losses arising from changing credit spreads.

Cross-border lending risk Cross-border lending risk is a legal risk that arises when cover pool assets are located in another country (different from the one the issuing bank is located in) and, therefore, subject to local bankruptcy laws that may not recognize the covered bond creditor's priority claim.

Cross-collateralization Cross-collateralization means that collateral securing one pool (or loan) is also used as collateral for another pool (or loan).

Currency risk Risk of losses due to adverse exchange rate movements.

DeckRegV Cover Register Statutory Order ("Deckungsregisterverordnung"), which regulates the content and the handling of the cover register.

Early redemption risk Risk that the Pfandbrief is repaid before its contractual maturity date.

ECBC European Covered Bond Council.

Eligibility criteria In order to be eligible for cover, assets must fulfil certain criteria which mainly depend on the Pfandbrief type.

Excess cover Cover requirement, saying that the net present value of the cover pool must exceed the net present value of the liabilities by at least 2 %, even under certain interest rate and currency stress scenarios.

Extension risk Risk that the Pfandbrief is repaid after its contractual maturity.

FlugBelWertV Regulation on the Determination of the Mortgage Lending Value of Aircraft ("Flugzeugbeleihungswertermittlungsverordnung").

Further cover assets To increase cover pool liquidity without changing the basic characteristics of the respective cover pools, the PfandBG allows—to a certain extent—the inclusion of further cover assets (as opposed to ordinary cover assets).

Gewährträgerhaftung German state guarantee scheme for public sector credit institutions (such as German savings banks and Landesbanks), which was abolished in July 2001. Debt issued in an interims phase from July 2001 to 2005 still benefits from the guarantee if it matures until the end of 2015. (Source: Moser et al. 2002)

Hard bullet maturity In structures with hard bullet maturities (as opposed to soft bullet maturities), the expected maturity date cannot be extended.

Insolvency administrator Administrator of the issuer's general insolvency estate.

Interest rate risk Risk of losses due to adverse interest rate movements.

Issuer default risk Risk of losses arising from the issuer's inability to fulfil his payment obligations in a full and timely manner.

Issuer risks Risks related to the issuer, such as issuer default risk, risks arising from cover pool management or further Pfandbrief issuance, and other issuer-related risks.

Jumbo Pfandbriefe Jumbo Pfandbriefe, also called Jumbos, fulfil a defined range of minimum standards: They have an issue size of at least EUR 1 bn (if this is not fulfilled with the initial issue, the issue size can be increased by tap issuance), annual fix rate coupons paid in arrears, and a hard bullet redemption. There are also requirements regarding stock market listing, syndicate banks, quoting, publishing of average spreads, transfer and buyback, loss of Jumbo Pfandbrief status and rules of good conduct for issuers and syndicate banks. Due to their large volume, high standardization and the market making requirements, Jumbo Pfandbriefe are considered to be very liquid. (Source: vdp 2013a).

KWG The German Banking Act ("Kreditwesengesetz").

Liquidity buffer Cover requirement, saying that the maximum cumulative liquidity gap within the next 180 days must be covered by liquid assets.

Mandatory overcollateralization Overcollateralization resulting from the cover requirements stipulated by the PfandBG.

Market value risk Risk of losses due to changes in realizable market values. This includes market risks (such as interest rate risk, currency risk, credit spread risk) and consequences of a liquidity shortfall (i.e. the risk of losses due to sale discounts that are caused by a lack of liquidity).

Matching cover calculation Daily calculations performed by the Pfandbriefbank to ensure that the required cover is always given.

Mortgage lending value Property value addressed by the PfandBG and specified in detail by the BelWertV. The mortgage lending value is based on permanent and sustainable features and ignores temporary and speculative elements.

Mortgage Pfandbrief A Pfandbrief backed by mortgages on real estate properties.

Natural hedging Reduction of risk exposure by altering asset and/or liability characteristics, as opposed to hedging by entering into a derivative contract.

Negative carry The bank's cost of financing an asset exceeds the yield earned on it.

Nominal cover Cover requirement, saying that the nominal of outstanding Pfandbriefe must be covered by a cover pool of at least the same nominal amount.

Operational risk Risk of losses arising from daily business activities, such as peoples' actions, systems, inadequate processes or external events. Operational risk can refer to the issuer or to other parties, such as the the cover pool monitor or the cover pool administrator.

Ordinary cover assets Assets that fulfil the Pfandbrief-type specific eligibility criteria.

Pari passu Two creditors that rank pari passu have the same rights and privileges.

Pass-through structure In pass-through structures, the investors are repaid from the natural amortization of the assets underlying the structure.

PfandBarwertV The Net Present Value Regulation ("Pfandbrief-Barwertverordnung") specifies the calculation of the net present value to be used for the determination of cover requirement.

PfandBG German Pfandbrief Act ("Pfandbriefgesetz").

Pfandbrief A covered bond under German Pfandbrief legislation.

Pfandbrief bank A bank holding a license for Pfandbrief issuance.

Pfandbrief issuer See Pfandbrief bank.

Pfandbrief legislation All legislative materials that are relevant in the context of German Pfandbriefe.

Political risk Risk of losses due to political changes or instability, which includes events such as war, revolutions and embargos.

Preferred creditors See privileged creditor.

Prepayment risk Risk of early repayment of cover pool debt.

Privileged creditors Pfandbrief holders and counterparties from cover pool derivatives.

Public Pfandbrief A Pfandbrief backed by claims against public sector debtors.

Real estate risk Risk of losses due to changing real estate prices.

Refinancing register The refinancing register ensures the insolvency remoteness and segregation of charges over real property held in a fiduciary capacity and facilitates refinancing through Pfandbriefe and securitization.

Refinancing risk Risk arising from maturity mismatches between assets and liabilities when the natural amortization of the cover pool is not sufficient to match the maturity profile of the outstanding Pfandbriefe.

RefiRegV The Funding Register Statutory Order ("Refinanzierungsregisterverordnung") regulates the content and the handling of the refinancing register according to § 22a–§ 22o of the German Banking Act.

Regulatory risk Risk that (unexpected) changes to the regulatory framework affect the Pfandbrief holder in a negative way.

Reinvestment risk Reinvestment risk refers to the uncertainty regarding the conditions to which cover pool proceeds can be reinvested. It arises from maturity mismatches when assets are being repaid earlier than Pfandbriefe, from asset prepayments or from defaults with recoveries.

Reputational risk Risk of losses arising from an adverse market perception of the issuer or the whole Pfandbrief segment.

Ring-fencing Separation of cover pool assets from the issuer's general insolvency estate.

Risk of timely Pfandbrief repayment Risk that the Pfandbrief is repaid earlier (early redemption risk) or later (extension risk) than the contractual maturity date.

SARA clause The "Selected Assets Required Amount" (SARA) clause prevents the excessive use of overcollateralization for the repayment of earlier maturing covered bonds to the disadvantage of later maturing bonds by allocating overcollateralization to the outstanding covered bonds on a pro-rata basis.

SchiffsBelWertV Regulation on the Determination of the Mortgage Lending Value of Ships and Ships under Construction ("Schiffsbeleihungswertermittlungsverordnung").

Segregation risk Risk of problems occurring in the context of cover pool separation upon issuer insolvency. Segregation risk comprises risks such as enforceability of voluntary overcollateralization, continuation of cover pool derivatives, cross-border lending risk, commingling risk, claw-back risk and set-off risk.

Set-off risk Risk that creditors of the issuer's general insolvency estate attempt to seize cover pool assets to set off their claims against the Pfandbrief issuer.

Settlement risk Risk of losses arising from the counterparty's failure to settle its obligations.

Ship Pfandbrief A Pfandbrief backed by mortgages on ships.

Soft bullet maturity Option to extend the original maturity by a pre-defined period without triggering a default event. During the extension period, a pre-specified coupon is paid.

Sovereign (credit) risk Risk that a government fails to fulfil its payment obligations in a full and timely manner.

Specialist bank principle Under the specialist bank principle (as opposed to the universal bank principle), the Pfandbrief bank is allowed to engage only in a restricted number of business activities, such as mortgage lending and/or public sector lending.

Strategic risk Risk of losses arising from strategic business decisions of the issuer.

Structural and legal risk Uncertainty regarding (i) the effectiveness of the mechanics in place to ensure bankruptcy remoteness and cover pool continuation, and (ii) the decision of cover pool liquidation.

Substitution risk Risk arising from the fact that the issuer can change the composition of the cover pool at any time by adding or removing cover pool assets, or by entering into new derivative contracts.

Time subordination In case of cover pool insolvency, the PfandBG does not restrict the amount of assets that can be liquidated at once for the repayment of a specific Pfandbrief maturity.

Traditional Pfandbriefe The main difference between traditional Pfandbriefe and Jumbo Pfandbriefe is the way they are structured. As opposed to Jumbo Pfandbriefe, traditional Pfandbriefe do not need to have a minimum issuance size of EUR 1 bn, and they can be issued as registered Pfandbriefe, not only in bearer form (as is the case for Jumbos). Traditional Pfandbriefe are typically not standardized but structured to fit the investor's need. (Source: vdp 2013b)

Transfer risk Risk that a government does not allow monetary transactions to take place with counterparties in other countries.

Universal bank principle Under the universal bank principle (as opposed to the specialist bank principle), there are no specific restrictions regarding the business activities a Pfandbrief issuer is allowed to engage in, apart from the general regulations of the German Banking Act.

vdp Association of German Pfandbrief Banks ("Verband Deutscher Pfandbriefbanken").

vdp credit quality differentiation model The vdp's credit quality differentiation model goes beyond legal requirements for matching cover calculations and accounts for differences in the sovereigns' creditworthiness by applying maturity-independent haircuts to the nominal value of public sector claims from European countries where the corresponding central state has a non-investment grade rating. Since end of February 2013, vdp member banks apply these standards on a voluntary basis.

Voluntary overcollateralization Overcollateralization exceeding the mandatory overcollateralization.

MIX
Papier aus verantwortungsvollen Quellen
Paper from responsible sources
FSC® C105338

If you have any concerns about our products,
you can contact us on
ProductSafety@springernature.com

In case Publisher is established outside the EU,
the EU authorized representative is:
Springer Nature Customer Service Center GmbH
Europaplatz 3, 69115 Heidelberg, Germany

Printed by Libri Plureos GmbH
in Hamburg, Germany